C000131697

"This book is a treasure-trove for Chı   nd theological bearings on questions of f   :se positions and offer helpful ways of adᴅᴜᴇssᴜᴏ   ᴇse matters. Above all, this is a book on creation that fills one's mind and heart with praise to the Creator!"

Matthew Levering, James N. and Mary D. Perry Jr. Chair of Theology at Mundelein Seminary

"The topic of creation has been the subject of a great deal of recent scholarship in Christian theology. This excellent, diverse collection of essays brings fresh new light to the topic not only by addressing challenging issues, but by bringing light to important aspects of the discussion that have often been ignored. The volume is refreshingly characterized by intellectual rigor, a passion for orthodoxy, and a deeply pastoral tone."

Michael Murray, senior visiting scholar at Franklin and Marshall College

"Creation is vast, the universe an incomprehensible diversity—'worlds without end.' This much is familiar, but who knew the *doctrine* of creation was equally far reaching? The essays in *Creation and Doxology* range far and wide, as do their authors' disciplines, and, while the question of origins is ably represented, the real surprise is the wide array of topics these chapters cover: everything from genes to Genesis, time and truth, matter and medicine. The doctrine of creation looms large over all areas of life. Of the many important takeaways in this book, one is surely the call to pastor theologians to tear down the dichotomy between the spiritual and the material. These essays remind us that the gospel is good news for the whole creation."

Kevin J. Vanhoozer, research professor of systematic theology at Trinity Evangelical Divinity School

CREATION AND DOXOLOGY

THE BEGINNING *and* END *of* GOD'S GOOD WORLD

EDITED BY GERALD HIESTAND
& TODD WILSON

IVP Academic

An imprint of InterVarsity Press
Downers Grove, Illinois

InterVarsity Press
P.O. Box 1400, Downers Grove, IL 60515-1426
ivpress.com
email@ivpress.com

©2018 by Gerald L. Hiestand and Todd A. Wilson

All rights reserved. No part of this book may be reproduced in any form without written permission from InterVarsity Press.

InterVarsity Press® is the book-publishing division of InterVarsity Christian Fellowship/USA®, a movement of students and faculty active on campus at hundreds of universities, colleges, and schools of nursing in the United States of America, and a member movement of the International Fellowship of Evangelical Students. For information about local and regional activities, visit intervarsity.org.

All Scripture quotations, unless otherwise indicated, are taken from The Holy Bible, New International Version®, NIV®. Copyright © 1973, 1978, 1984, 2011 by Biblica, Inc.™ Used by permission of Zondervan. All rights reserved worldwide. www.zondervan.com. The "NIV" and "New International Version" are trademarks registered in the United States Patent and Trademark Office by Biblica, Inc.™

Cover design: David Fassett
Interior design: Beth McGill
Image: © The Creation, from Luther Bible, German School / Bible Society, London, UK / Bridgeman Images

ISBN 978-0-8308-5386-1 (print)
ISBN 978-0-8308-7403-3 (digital)

Printed in the United States of America ∞

InterVarsity Press is committed to ecological stewardship and to the conservation of natural resources in all our operations. This book was printed using sustainably sourced paper.

Library of Congress Cataloging-in-Publication Data

Names: Hiestand, Gerald, 1974- editor.
Title: Creation and doxology : the beginning and end of God's good world / Gerald Hiestand and Todd Wilson, eds.
Description: Downers Grove : InterVarsity Press, 2018. | Series: Center for Pastor Theologians (Series) | Includes index.
Identifiers: LCCN 2018022667 (print) | LCCN 2018029823 (ebook) | ISBN 9780830874033 (eBook) | ISBN 9780830853861 (pbk. : alk. paper)
Subjects: LCSH: Creation.
Classification: LCC BS651 (ebook) | LCC BS651 .C69155 2018 (print) | DDC 231.7/65—dc23
LC record available at https://lccn.loc.gov/2018022667

| P | 23 | 22 | 21 | 20 | 19 | 18 | 17 | 16 | 15 | 14 | 13 | 12 | 11 | 10 | 9 | 8 | 7 | 6 | 5 | 4 | 3 | 2 | 1 |
| Y | 37 | 36 | 35 | 34 | 33 | 32 | 31 | 30 | 29 | 28 | 27 | 26 | 25 | 24 | 23 | 22 | 21 | 20 | 19 | 18 |

To the Fellows of the St. Anselm Fellowship,

who have been with us since the beginning.

Contents

Acknowledgments

A S WITH THE PREVIOUS VOLUMES in our Center for Pastor Theologians conference series, we are especially grateful to the men and women who served as presenters at the conference and who have contributed to the present volume. These essays are gracious and clear, demonstrating a depth of both pastoral and theological insights. We are grateful to partner with such an excellent group of ecclesial theologians, academic theologians, scientists, cultural critics, and Christian leaders.

We likewise owe a debt of gratitude to the Center for Pastor Theologians (CPT), the organizer of the conference from which the papers of this book are drawn. The Center continues to serve as a catalyst for our work and has been a repository of wisdom and counsel on all things pastoral and theological. The other members of the board of the Center (John Yates, Michael LeFebvre, and John Isch), as well as the staff (Jeremy Mann and Zach Wagner), deserve our gratitude and bear a measure of responsibility for any blessing this book brings to the church. A special thanks to Zach for indexing this volume and organizing the many details of the conference.

Likewise, we are profoundly grateful for Calvary Memorial Church in Oak Park, Illinois, the congregation where we are privileged to minister. Calvary has graciously served as the home for the CPT for the better part of a decade, and it is not an understatement to say that the CPT would not be what it is without Calvary's partnership and support.

We are thankful for IVP Academic and their commitment to ecclesial theology and the CPT's vision of the pastor theologian. Our editor, David McNutt, deserves a special word of thanks; his enthusiastic participation in the production of this book has gone a long way toward making it a reality.

We are deeply grateful for the partnership of the CPT's four senior theological mentors: Scott Hafemann, Doug Sweeney, Paul House, and Kevin Vanhoozer. Their commitment to the CPT's mission, their contribution to our Ecclesial Theologian Fellowships, and their friendship and encouragement have been an important catalyst for the CPT project.

Finally, to our families, especially our wives, we remain ever grateful. Their patient endurance for projects like this one, in the midst of our already busy schedules, is a gift that we do not take lightly. May the Lord pay them back tenfold what they have given to us!

Introduction

In Praise of Beauty: The Native Connection Between Creation and Doxology

GERALD HIESTAND
AND TODD WILSON

The world is charged with the grandeur of God.

GERARD MANLEY HOPKINS,
"GOD'S GRANDEUR"

The lovely things kept me far from you.

SAINT AUGUSTINE, *CONFESSIONS*

PLATO ONCE ASKED, "Isn't this dreaming: whether asleep or awake, to think that a likeness is not a likeness but rather the thing itself that it is like? But someone who, to take the opposite case, believes in the beautiful itself, can see both it and the things that participates in it. . . . He is very much awake."[1]

Even the pagans get it right sometimes. Plato rightly saw that beautiful things are beautiful precisely because they participate in Beauty itself. Of course,

---

[1]Plato, *Republic* 5.476.c, in *Plato: The Complete Works*, ed. John M. Cooper, trans. G. M. A. Grube and C. D. C. Reeve (Indianapolis: Hackett, 1997), 1103.

not everyone sees beyond the beautiful things to Beauty itself. Such people, Plato tells us, are asleep to reality. The "awake ones" are those who properly recognize that the beautiful things are but shades of a higher and truer Good.

In his own limited and prescient way Plato is affirming Saint Paul's seminal insight (found in the opening chapter of Romans) that humanity has fallen under the judgment of God because it has severed the connection between the Creator and the creation. The created world reveals the beauty, power, and glory of the uncreated God. But humanity has confused the beautiful things with Beauty itself. We have chosen to live our lives willfully asleep to the reality of God; we have fallen in love with the beautiful things and have abandoned the Beautiful One. In a deep and tragic irony, the very things that were intended to point us to God have obscured our knowledge of him. The beautiful things are blessings when we receive them with thanks. But they are false gods when we worship them in place of the Creator. We have made the means an end, and the beautiful things, rather than leading us to God, have led us only to ourselves.

It is easy to see why humanity is so easily seduced by the beautiful things. The beautiful things make no demands on us. They are gods that we can control, that bow before us. But the Beautiful One transcends us. He is not at our beck and call, bending himself to our will. Beauty, in the Person of God himself, calls us to allegiance and submission. When we acknowledge the existence of the Beautiful One, we are compelled to acknowledge that we are mere creatures, finite, subordinate. Beauty calls us to acknowledge, in our recognition of God as Creator, that we are beautiful only insofar as we surrender ourselves to one who is Beauty himself.

This is why Paul will go on to state that a posture of thanksgiving renders idolatry nearly impossible. To give genuine thanks for creation is to acknowledge that there is One above and beyond humanity who has given it. To give thanks for the world and our very selves necessarily compels us to acknowledge that the Lord *is*, and that he is *good*, and that he *gives*. It reminds us that we ourselves are not the good God, but that we stand in a posture of humility and need—that we are recipients of grace. Thankfulness rightly orders human self-understanding with respect to the creation of which we are a part, and with respect to the God who made and gave it to us. This is why a refusal to give thanks to God for the good world he has given and a refusal to acknowledge the iconic nature of creation go hand in hand. To thankfully acknowledge creation as a beautiful *gift* is to acknowledge that there is necessarily a Beautiful *Giver*. At its core, thankfulness establishes the relationship between the gift and the giver. To quote another pagan who also got it right, "When you look at the gift, look at the giver too."[2] It is impossible to give genuine praise to God for the good things of the world and idolize these things at the same time.

Saint Augustine understood the need to acknowledge the goodness of God in the goodness of his creation: "If physical objects give you pleasure, praise God for them and return love to their Maker lest, in the things that please you, you displease him. . . . For all that comes from him is unjustly loved if he has been abandoned."[3] In the same spirit, this collection of essays is an effort to "return love" to the Maker of the world, to acknowledge his ultimate transcendence in all things and before all things, to give him thanks, and to affirm that praise to the Creator is the ultimate telos of creation. Toward that end, the essays in this book seek explicitly to establish and celebrate the native connection between creation and doxology, between the beautiful things that have been made and the Beautiful One himself, between the created things and the Creator God who gave them.

Some of the essays in this volume appropriately wade into the intramural debates still being waged regarding Christianity's posture toward post-Darwinian science. And some of the essays draw out the ethical and pastoral implications that necessarily flow from a robust, biblical doctrine of creation. In a day when (too) much Christian theological reflection on the doctrine of creation has been preoccupied with apologetic discussions and in-house debates regarding how to read Genesis, there is a need for theologians—both pastoral and academic—to be reminded that creation is first and foremost an occasion for praise and thanksgiving. To miss this aspect of the doctrine of creation is to miss its central node.

The essays are drawn from the papers presented at the 2017 annual theology conference of the Center for Pastor Theologians. The conference

---

[2]Seneca, *Thyestes* 416, in *Seneca: Six Tragedies,* trans. Emily Wilson (Oxford: Oxford University Press, 2010).

[3]Augustine, *Confessions* 4.12, in Augustine, *Confessions*, trans. Henry Chadwick (Oxford: Oxford University Press, 1992), 18.

brought together nearly three hundred pastors, academics, students, and lay leaders for an invigorating discussion about the relevance and import of the doctrine of creation. The spirit of the conference was, as in past years, both irenic and engaging. As is evident from the essays here, not all the contributors agree on every aspect of the doctrine of creation. Some are less persuaded than others that the claims of post-Darwinian science can be easily reconciled to the core narrative of the Christian faith. Others are more optimistic. But all the contributors are equally persuaded that, however one might think about the question of origins, the proper posture of the creature before the Creator is that of praise and thanksgiving.

The glory of the gospel is that when we as mere creatures gratefully embrace our creaturely status, the Creator remarkably, beyond hope or expectation, makes us more than mere creatures. He does this in a way that stretches beyond the philosophy of the Greeks and the prophecies of the Jews—by becoming his own creation. Thus Saint Irenaeus speaks for the fathers of the church when he states the wondrous mystery of the incarnation and our redemption: "Our Lord Jesus Christ did, through transcendent love, become what we are, that he might bring us to be even what he himself is."[4] He became as us, mere creatures, that we might become as him, true children of God.

Our prayer is that this volume will not only deepen the reader's understanding of a central doctrine of the Christian faith but also, more importantly, deepen the reader's love for God and foster a genuine and humble posture of thankfulness for all that God has done in gracing us, and our world, with himself.

We invite the reader to exclaim with Paul, "Thanks be to God for his indescribable gift!" (2 Cor 9:15).

---

[4]Irenaeus, *Against Heresies*, ed. Alexander Roberts and James Donaldson, trans. Cleveland A. Coxe, in *Ante-Nicene Fathers*, vol. 1, *The Apostolic Fathers with Justin Martyr and Irenaeus* (repr., Peabody, MA: Hendrickson, 2004), book 5, preface.

Part One

The DOCTRINE of CREATION EXPRESSED

1

Reading Genesis 1 with the Fourth Commandment

The Creation Week as a Calendar Narrative

MICHAEL LeFEBVRE

O NE OF THE EARLIEST COMMENTARIES on the creation week is the fourth commandment:[1] "Six days you shall labor, . . . but the seventh is a Sabbath to the LORD your God. . . . For in six days the LORD made heaven and earth, . . . and rested on the seventh day" (Ex 20:9-11).[2] This commandment interprets the creation week as a pattern for Israel's labor and rest.

In recent decades, attention has focused on the creation week's historical character. Is it descriptive history, describing how creation actually happened? Or is it poetic? How does the creation narrative relate to the narrative told by evolutionary science? These discussions tend to focus on the six days.[3] But if there is one point of consensus through history, it is the text's primary concern with the seventh day as enshrined in the fourth commandment.

---

[1]The fourth commandment in Jewish, Greek Orthodox, and Protestant numbering is the third commandment in Roman Catholic and Lutheran traditions.

[2]Scripture quotations are from the ESV.

[3]Ken Ham, *The Lie: Evolution/Millions of Years* (Green Forest, AR: Master Books, 2013), only mentions the Sabbath once (184) to show the twenty-four-hour nature of the other six creation days. Hugh Ross, *A Matter of Days: Resolving a Creation Controversy* (Corvina, CA: Reasons to Believe, 2015), only mentions the seventh day (which lacks the "evening and morning" marker) to support the six days as long periods of time (73-75, 83-86, 242). Cf. Robert Godfrey, *God's Pattern for Creation: A Covenantal Reading of Genesis 1* (Phillipsburg, NJ: P&R, 2003), 59.

The Sabbath was not the only holy day in Israel. Israel had numerous festivals, many of which have associated narratives. How might the other calendar narratives in the Pentateuch offer insights to help assess the creation week? This essay will explore this question, beginning with the calendar employed in the Genesis flood narrative.[4]

DATES IN THE FLOOD NARRATIVE

Flood stories were widespread in the ancient world. One distinctive of the biblical flood account is its use of dates. There are five dates in the Genesis flood narrative. This is remarkable, since those are the only dates in the entire book of Genesis.

Typically in ancient literature, an event's timing was indicated by relating it to another event, not by using dates. *Timeline dating*—plotting events on a transcendent timeline with dates—is common today, but ancient texts used *event sequencing*, temporally marking an event by relating it to other events.[5] Note the following examples in Genesis: "To Seth also a son was born. . . . At that time people began to call upon the name of the LORD" (Gen 4:26); "When man began to multiply on the face of the land . . . , the sons of God saw that the daughters of man were attractive" (Gen 6:1-2); "After these things the word of the LORD came to Abram" (Gen 15:1). Throughout Genesis, event sequencing is used. But five dates appear in the flood narrative (and nowhere else in the entire book of Genesis):

1. "In the six hundredth year of Noah's life, in the second month, on the seventeenth day of the month, on that day all the fountains of the great deep burst forth" (Gen 7:11).

2. "In the seventh month, on the seventeenth day of the month, the ark came to rest on the mountains of Ararat" (Gen 8:4).

3. "And the waters continued to abate until the tenth month; in the tenth month, on the first day of the month, the tops of the mountains were seen" (Gen 8:5).

---

[4]This essay summarizes my larger study of the relationship between calendars and creation in light of Genesis 1:1–2:3.

[5]Sacha Stern, *Time and Process in Ancient Judaism* (Oxford: Littman Library of Jewish Civilization, 2007), 12; Norbert Elias, *Time: An Essay* (Oxford: Blackwell, 1992), 106-7; Stern, *Time and Process*, 91-102.

4. "In the six hundred and first year, in the first month, the first day of the month, the waters were dried from off the earth" (Gen 8:13).

5. "In the second month, on the twenty-seventh day of the month, the earth had dried out" (Gen 8:14).

Table 1. Dates in the flood narrative

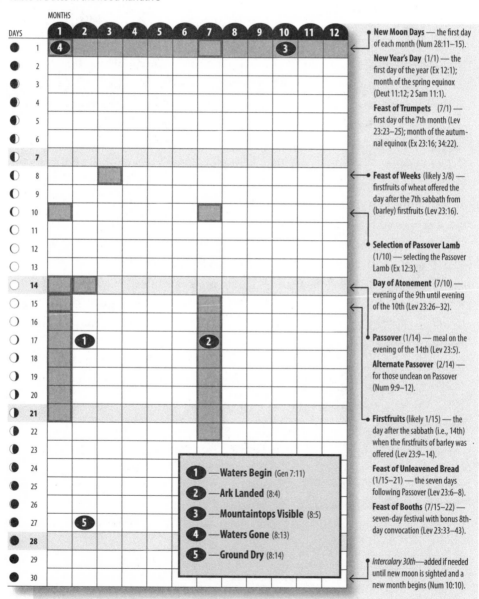

An important insight emerges when these dates are plotted against the festival calendar of Israel (see table 1). Three of the five fall directly on Mosaic festival dates. The only exceptions are the first and last, which nonetheless fall at the midpoint of Israel's grain-harvest festivals. All five dates appear to be "scheduled" with reference to Israel's festivals. A survey of each date illuminates this relationship.

1. *The beginning of the flood (Gen 7:11).* The flood's beginning date (2/17) is at the center of Israel's grain festivals. The early spring festivals—Passover, Unleavened Bread, and Firstfruits (1/14–21)—began the barley harvest. The Feast of Weeks in late spring marked the wheat harvest. During a good harvest year in Israel, the rains ended by springtime.[6] A thunderstorm during the grain harvest endangered crops and was regarded as a sign of judgment, as illustrated by the words of Samuel: "Is it not wheat harvest today? I will call upon the LORD, that he may send thunder and rain. And you shall know and see that your wickedness is great" (1 Sam 12:17; cf. Ex 9:31-32; Prov 26:1). Dating the start of the deluge in the middle of Israel's grain harvest adds to its ominous character.

2. *The ark's landing (Gen 8:4).* "The ark came to rest on the mountains of Ararat" on the seventeenth day of the seventh month (Gen 8:4). In later Israel, this date would fall during the Feast of Booths. Moses appointed that festival to commemorate Israel's safe passage through the wilderness to the Promised Land (Lev 23:39-43; Num 2:1-34). Similarly, Noah's date marked his safe journey through a watery "wilderness," arriving at Mount Ararat (cf. 1 Kings 6:1).

3. *When mountaintops became visible (Gen 8:5).* Noah had his first sight of land on 10/1, three months after the ark's landing (7/17) and three months before the waters were gone (1/1). At that point, Noah saw the mountaintops and sent out birds "to see if the waters had subsided" (Gen 8:8) and whether foliage was growing again (Gen 8:11). In later Israel, that same date was a new-moon day in between Israel's festival years. The previous festival year ended with the Feast of Booths (7/15–22), and the next began on New Year's Day (1/1). The interim was Israel's rainy season, when Hebrew farmers planted for the new year and watched to see whether God would give them bounty

---

[6]David Hopkins, *The Highlands of Canaan: Agricultural Life in the Early Iron Age,* The Social World of Biblical Antiquity Series 3 (Decatur, GA: Almond Press, 1985), 87. Cf. Deut 11:14; Jer 5:24; Joel 2:23-24.

the next year. Noah's hopeful glimpse of land and the plucking of its first leaves fit well with Israel's experience at that same season.

4. When the waters were gone (Gen 8:13). By New Year's Day (1/1) the waters were gone. New Year's Day is a natural "new beginnings" point. By highlighting this date for the end of the flood, later Israelites would celebrate the new year remembering how Noah "removed the covering of the ark and looked" (Gen 8:13) and saw a new beginning granted in God's grace.

5. When the ground was dry (Gen 8:14). The ground was completely dry on 2/27. The significance of the flood's beginning in the heart of the grain harvest has already been noted. The same applies to its conclusion on a date one year and an even ten days after. If the storm's beginning during the harvest was a sign of judgment on that year's plantings, the restoration of dry ground during the harvest marked the return of normal agricultural order and bounty.

These correspondences suggest that the alignment between the five dated flood events and later Israel's festival calendar is not coincidental. Noah's flood was retold in a manner that related his "exodus" to Israel's festival worship and agricultural labors. If this reading is correct, one might still ask whether Noah's flood actually took place along these dates, or whether these dates were added anachronistically. One further feature indicates these are not dates recorded from observation but are a literary construction: the flood narrative uses schematic, thirty-day months rather than actual varying-length months. This is prima facie evidence of a constructed (rather than observed) timeline.

In ancient lands, the length of a month was based on lunar observations. The old month (lit., "moon," *ḥōdeš*) continued until the first sliver of the next moon appeared. That sighting marked the first day of the new month/ new moon.[7] Actual months therefore varied in length, roughly evenly, between twenty-nine or thirty days.[8] However, this uncertainty posed a problem for drafting legal texts or making economic projections. Therefore, "a schematic 360-day year . . . [of twelve] consecutive 30-day months" was used for economic calculations and legal texts.[9] Descriptive texts based on

---

[7]Francesca Rochberg-Halton, "Calendars, Ancient Near East," in *Anchor Bible Dictionary*, ed. David Noel Freedman (New York: Doubleday, 1992), 1:810.

[8]Peter J. Huber, *Astronomical Dating of Babylon I and Ur III*, Occasional Papers on the Near East 1/4 (Malibu, CA: Udena, 1982), 24-25.

[9]Jonathan Ben-Dov, "Calendars and Festivals," in *The Oxford Encyclopedia of the Bible and Law*, ed. Brent A. Strawn (Oxford: Oxford University Press, 2015), 1:88.

observations reported months varying between twenty-nine and thirty days. But legal texts composed for future instruction invoked schematic, thirty-day months.

The flood narrative uses schematic, thirty-day months. The five months between the beginning of the flood (on 2/17) and the ark's resting on Mount Ararat (on 7/17) are rendered as 150 days (Gen 7:24; 8:3), being five months of thirty days each. Two or three of those months would have been twenty-nine days in length if observational data were employed, giving a count of 147 or 148 days. A length of 150 days would not be possible. The flood dates, therefore, cannot be based on observations but have the form of a legal construction.

This conclusion is not to suggest a problem in the Genesis account. On the contrary, it is hard to imagine the author would overlook such a simple calculation if an appearance of journalistic description were intended. This conclusion indicates that we are dealing with a legal text by design, rather than an observational record. The use of aesthetically balanced dates and numbers throughout the passage, such as 7s, 10s, 40s, 150, along with the use of schematic months, indicates the constructed nature of this narrative's dates for a legal (rather than journalistic) purpose. It is therefore proposed that the flood account is an agricultural and festival calendar in narrative form: a *calendar narrative.*

This function for the flood narrative is comparable to the contemporary practice of telling Jesus' birth story on December 25. Churches do so, not to assert that Jesus was actually born on that date, but to inform Christian observances on that date. Similarly, the flood narrative re-maps the events of Noah's deluge to the calendar of later Israel's agricultural labors and harvest festivals for its instructional value. The plausibility of this argument is strengthened when the same features are noted in the Pentateuch's exodus narratives.

DATES IN THE EXODUS NARRATIVE

Like Genesis, Exodus uses event sequencing, not timeline dating. No dates occur in Exodus until Passover night (Ex 12). Then five dates suddenly appear in rapid succession (Ex 12:1–13:16), and eleven more follow at later points of Israel's journey (in Exodus–Deuteronomy). Like the flood dates, these sixteen exodus dates align with key dates on Israel's festival calendar (see table 2).

Table 2. Dates in the exodus narrative

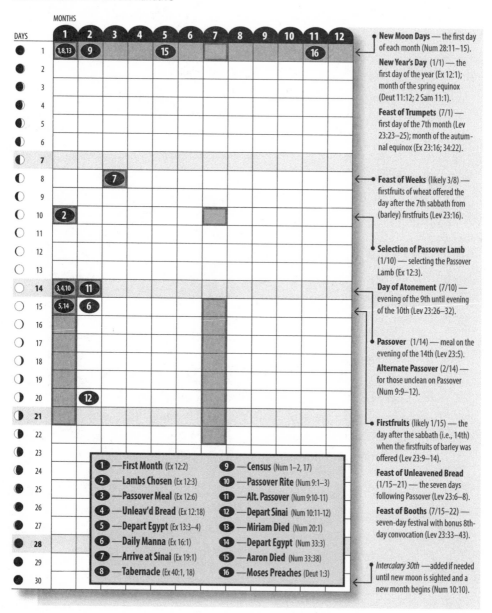

New Moon Days — the first day of each month (Num 28:11–15).

New Year's Day (1/1) — the first day of the year (Ex 12:1); month of the spring equinox (Deut 11:12; 2 Sam 11:1).

Feast of Trumpets (7/1) — first day of the 7th month (Lev 23:23–25); month of the autumnal equinox (Ex 23:16; 34:22).

Feast of Weeks (likely 3/8) — firstfruits of wheat offered the day after the 7th sabbath from (barley) firstfruits (Lev 23:16).

Selection of Passover Lamb (1/10) — selecting the Passover Lamb (Ex 12:3).

Day of Atonement (7/10) — evening of the 9th until evening of the 10th (Lev 23:26–32).

Passover (1/14) — meal on the evening of the 14th (Lev 23:5).

Alternate Passover (2/14) — for those unclean on Passover (Num 9:9–12).

Firstfruits (likely 1/15) — the day after the sabbath (i.e., 14th) when the firstfruits of barley was offered (Lev 23:9–14).

Feast of Unleavened Bread (1/15–21) — the seven days following Passover (Lev 23:6–8).

Feast of Booths (7/15–22) — seven-day festival with bonus 8th-day convocation (Lev 23:33–43).

Intercalary 30th — added if needed until new moon is sighted and a new month begins (Num 10:10).

Legend:
1 — First Month (Ex 12:2)
2 — Lambs Chosen (Ex 13:3)
3 — Passover Meal (Ex 12:6)
4 — Unleav'd Bread (Ex 12:18)
5 — Depart Egypt (Ex 13:3–4)
6 — Daily Manna (Ex 16:1)
7 — Arrive at Sinai (Ex 19:1)
8 — Tabernacle (Ex 40:1, 18)
9 — Census (Num 1–2, 17)
10 — Passover Rite (Num 9:1–3)
11 — Alt. Passover (Num 9:10-11)
12 — Depart Sinai (Num 10:11-12)
13 — Miriam Died (Num 20:1)
14 — Depart Egypt (Num 33:3)
15 — Aaron Died (Num 33:38)
16 — Moses Preaches (Deut 1:3)

A full examination of these exodus dates and their festival correspondences is not possible in this essay, although such an assessment has
been provided elsewhere.[10] Since the exodus is the presenting narrative
of Israel's calendar, it should not be surprising to find significant correspondences. But a further question follows: Are the exodus narrative
dates based on the observed timing of original events, or are these dates
added to create festival alignments? Several lines of evidence point to
the latter conclusion.[11]

For example, Exodus repeatedly tells us that Sinai was a "three days'
journey" from Egypt (Ex 3:18; 5:3; 8:27). However, nearly three months
separate the date Israel is said to have left Egypt (Num 33:3) and the date of
their reported arrival at Mount Sinai (Ex 19:1). Various efforts have at-
tempted to resolve this discrepancy,[12] but the mathematical dilemma remains. The best explanation is that the compiler was not concerned to
smooth out chronological details. Instead, for liturgical purposes the journey's beginning was matched to the date of Israel's barley-harvest festival
and their arrival was matched to the date of Israel's wheat harvest. Exodus
takes Israel's "three days' journey" and maps it to these harvest dates for
worship instruction. Later Israel was taught to remember God's power to
bring their forefathers into the land, precisely at those times when they were
gathering their harvests in that land.

Many such chronological dilemmas have been noted in the exodus narrative. Jan van Goudoever cataloged several of them, concluding, "From
such conflicting indications it is clear that the 'calendar' in the Torah is not
consistent. There are either different traditions, which are not harmonized,
or some alterations were made by writers or redactors which disturbed a
'calendar' which was originally consistent."[13] But sloppy redaction is not the
only explanation for these idiosyncrasies. The best explanation is that these

---

[10]Michael LeFebvre, "The Liturgical Function of Dates in the Pentateuch," in *Exploring the Composition of the Pentateuch*, ed. L. S. Baker Jr. et al. (University Park: Pennsylvania State University
Press, forthcoming).

[11]Further evidence beyond the following example is discussed in LeFebvre, "Liturgical Function."

[12]E.g., John I. Durham, *Exodus*, Word Biblical Commentary (Waco, TX: Word, 1987), 40 (but
contra Durham, see Ex 3:12); Benno Jacob, *The Second Book of the Bible: Exodus*, trans. Walter
Jacob (Hoboken, NJ: Kravets, 1992), 125; W. H. Gispen, *Exodus*, trans. Ed van Derr Maas, Bible
Student's Commentary (Grand Rapids: Zondervan, 1982), 57.

[13]Jan van Goudoever, *Biblical Calendars* (Leiden: Brill, 1961), 56.

exodus dates are not intended to form a timeline. The actual events occurred on a timescale and perhaps in an order that is not preserved, and their recounting has been mapped to the harvest calendar of Israel in order to inform the people's seasonal labors and worship.

CALENDAR NARRATIVES IN THE TORAH

Like Hans Christian Anderson's "ugly duckling," these narratives are ungainly when presumed to give journalistic chronologies. But when approached as a form of legal instruction, these same dates bring out the text's beauty. The biblical title for the Pentateuch is "Torah" (Hebrew for "law") because it served as the law collection of ancient Israel. Not only the statutes but also the narratives of the Pentateuch served as Israel's foundational law. There is a legal purpose even for the narratives in the Torah.

Some of the Torah's narratives provide instruction about ritual practices, such as narratives about circumcision (Gen 17:1-14; cf. Ex 4:24-26) or certain dietary restrictions (Gen 32:22-32). Other Torah narratives provide instruction concerning the sanctity of various holy sites (e.g., Gen 22:1-14; 14:18-20; 28:10-22) and Israel's legal right to certain disputed territories (e.g., Gen 26:6-33; Num 21:24-30; cf. Judg 11:4-28). The legal function of narratives in the Pentateuch further emerges in stories that teach the sanctity of holy objects (Ex 26:1–40:38) or the divine sanction of certain offices, such as Ephraimite rule (Gen 48:1-22) or the primacy of Aaron's house among the Levites (Num 16:1–17:13).[14] Furthermore, many of the Torah's stories serve as narratives of specific statutes, such as Jacob's favoritism to the eldest of his second wife (Gen 29:31; 37:3) contrary to Deuteronomy 21:15-17.[15] In these and other ways, the narratives of the Torah participate in its legal instruction alongside the statutes.

Dates in the Torah's narratives participate in this broader legal function of the Pentateuch. Once the calendrical role of dated narratives within the Torah is recognized, this discovery offers insight for that most controversial of the Pentateuch's calendars: the creation week.

---

[14]Friedemann W. Golka, "The Aetiologies in the Old Testament: Part 1," *Vetus Testamentum* 26, no. 4 (1976): 410-28; Golka, "The Aetiologies in the Old Testament: Part 2," *Vetus Testamentum* 27, no. 1 (1977): 36-47; Jon D. Levenson, "The Theologies of Commandment in Biblical Israel," *Harvard Theological Review* 73 (1980): 17-33.

[15]Calum Carmichael, *The Laws of Deuteronomy* (Ithaca, NY: Cornell University Press, 1974), 61-62.

Genesis 1:1–2:3 as a Calendar Narrative

There are at least three calendrical features of the flood and exodus narratives that are also found in the creation week, suggesting all three date-laden narratives serve this calendrical purpose.

First, the creation week is structured around dates like the flood and exodus narratives. The creation week does not provide month dates like those other calendar narratives, but it does give week dates. Days of the Hebrew week were identified by number. A count of six days that culminates in a seventh-day Sabbath indicates week dates. Since the Pentateuch indicates temporality with event sequencing (not timeline dating), a narrative structured around dates is the first signal of a calendar.

Second, like the flood and exodus narratives, the creation week maps events to dates, even though the events do not fit the chronology of those dates. This can be illustrated in the following example.

Most scholars agree that the word *yôm* (day) in the creation week indicates normal, twenty-four-hour days (with the exception of one instance in v. 5, where "day" refers to the daylight half of a twenty-four-hour period in contrast with "night"). There are scholars who read the creation days as long eras of time,[16] but the repeated reference to evening and morning between days and their counting with ordinal numbers are strong indicators this is a normal week.[17] Some events within the creation week, however, do not fit within the confines of twenty-four-hour days. For example, day three reports, "And God said, 'Let the earth sprout [*dāšā'*] vegetation, plants yielding [*mazrîaʿ*] seed, and fruit trees bearing [*ʿāśāh*] fruit in which is their seed . . .' And it was so" (Gen 1:11). The verbs in this passage describe typical plant growth from sprouting (*dāšā'*) to fully mature plants and trees laden with (*ʿāśāh*) fruit. The passage does not say anything about instantaneous creation but describes a process of seeds sprouting into plants that takes months and even years to complete. But the author identifies this work with the third day without exhibiting any effort to indicate a rapid process that makes it fit chronologically. This is a feature we found in the other calendar narratives and points to a legal rather than journalistic use of dates. Genesis 1:1–2:3 contains numerous

---

[16]E.g., Ross, *Matter of Days*.

[17]Gordon Wenham, *Genesis 1–15*, Word Biblical Commentary 1 (Waco, TX: Word, 1987), 19.

chronological difficulties like this that have been frequently discussed through church history.[18] I would suggest that these features are another indication of a calendar narrative.

A third indication that we are dealing with a calendar narrative is the thematic alignment of specific events with the dates given. As Noah's arrival at Mount Ararat fits theologically with the Festival of Booths, the events of the creation week fit thematically with the weekly activities of the average Hebrew household both in its overall rhythms and in its day-one to day-seven progress.

In its rhythms, the week relates God's works to those of the common Hebrew laborer's week. The Creator rests not only on the Sabbath day but also each evening of the week. God is said to complete his daily labor during daylight hours and to cease from evening to morning, like a typical human worker (Ps 104:23). "The structure of the account," writes C. John Collins, "shows us that our author has presented God as if he were a craftsman going about his workweek."[19] Thus the creation week applies God's works to the practical cadences of Israel's households. The day-one to day-seven progress of events also relates to the daily concerns of the typical Hebrew family—namely, day-by-day progress in food production. Remarkably, the shaping theme of the creation week is food production. A closer look at the structure of the creation week will draw this out.

At the beginning of the narrative, the presenting problem to be solved is the land's barrenness: "the earth was without form and void [*tōhû wābōhû*]" (Gen 1:2). The first of these terms, *tōhû* (without form), refers to the world's need to be brought into an orderly condition. The second term, *bōhû* (void or empty), refers to the earth's barrenness resulting from its disordered condition.[20] This lack of fruitfulness is the presenting problem that the creation week labors to resolve, the resolution of which the Sabbath celebrates.

---

[18]G. W. Butterworth, *Origen on First Principles* (London: SPCK, 1936), 288; Robert Letham, "'In the Space of Six Days': The Days of Creation from Origen to the Westminster Assembly," *Westminster Theological Journal* 61 (1999): 149-74.

[19]C. John Collins, *Genesis 1–4: A Linguistic, Literary, and Theological Commentary* (Phillipsburg, NJ: P&R, 2006), 77. Cf. Origen in Butterworth, *First Principles*, 288.

[20]Iain Provan, *Discovering Genesis: Content, Interpretation, Reception* (Grand Rapids: Eerdmans, 2015), 62.

It is widely recognized that the six creation days are presented in two panels of three-day sets. The first triad (days one, two, and three) describes the ordering of realms into which the second triad (days four, five, and six) inserts residents.[21] This parallelism between the two sets is widely discussed.[22] Recently, Philippe Guillaume has refined this pattern, pointing out the emphasis on time in the opening days of both panels as well as in the climactic Sabbath. "The Sabbath and Gen 1," Guillaume writes, "are the creation of . . . a sacred calendar. . . . The first, fourth and seventh days are devoted exclusively to the creation of rhythms."[23] On the first day, the rhythm of day and night is created. The fourth day introduces the sun, moon, and stars "to separate the day from the night . . . [and to] be for signs and for seasons [lit., "festivals," *môʿadîm*], and for days and years" (Gen 1:14). The parallelism between the first days of each triad is calendrical. Then in the latter two days of each panel, the terrestrial order enabling fruitfulness (days two and three) and "fruit eaters" (days five and six) follows. This text is indeed a "sacred calendar" with an emphasis on the seasons and the order that enable human laborers to make the barren realms fruitful following the likeness of God, the Model Worker. That this principle of fruit-bearing is the focus of the calendar is further indicated by the teleological statements at the close of each three-day panel.

At the end of day three (and of the ordering of realms) the text reports, "And God said, 'Let the earth sprout vegetation, plants yielding seed, and fruit trees bearing fruit . . .' And it was so" (Gen 1:11). Then, at the end of day six (and of the populating of earth's realms) the text adds, "And God said, 'Behold, I have given you every plant yielding seed that is on the face of all the earth, and every tree with seed in its fruit. You shall have them for food. And to every beast of the earth and to every bird of the heavens . . . , I have given every green plant for food,'" (Gen 1:29-30). The first panel thus describes the order God instilled into the world leading to its fruitfulness. The second panel describes the denizens of the various realms, culminating in permission to eat of the world's fruitfulness.

---

[21]"Thus the heavens and the earth [realms] were finished, and all the host [residents] of them" (Gen 2:1).
[22]Hermann Gunkel, *Genesis*, Göttinger Handkommentar zum Alten Testament 1 (Göttingen: Vandenhoeck & Ruprecht, 1917), 118; J. G. von Herder, *Älteste Urkunde des Menschengeschlechts* (Tübingen: J. G. Cotta, 1806), 1:129-30.
[23]Philippe Guillaume, *Land and Calendar: The Priestly Document from Genesis 1 to Joshua 18*, Library of Hebrew Bible/Old Testament Studies 391 (New York: T&T Clark, 2009), 47.

Divine proclamations of goodness are woven through the passage in a manner that identifies these statements of fruitfulness as part of the week's structure and not just part of their respective days. The six days are marked by a sevenfold repetition of the announcement, "And God saw that it was good" (Gen 1:4, 10, 12, 18, 21, 25, 31). But these seven pronouncements are not placed evenly throughout the six days as one might expect. Aesthetically, it might seem attractive to place one of these statements on each of the six days, with the climactic seventh declaration added prior to the seventh day. However, day two is left without any mention of goodness. This appears to be done to retain a final count of seven proclamations while also allowing for a doubling of the announcement on days three and six, the final days of the two panels.

Thus, days three and six each begin with the work proper to that day, punctuated with a statement of its goodness. Then those same days each have an added statement indicating the world's status on that day in its progress from disorder and barrenness (*tōhû wābōhû*) to fruitfulness, followed by a second declaration of goodness (see table 3). This pattern confirms the design of the fruitfulness statements at the end of days three and six as part of the overall teleology of the week and not simply as events of those days.

Table 3. Structure of the creation week

	Realms ("heavens and earth," Genesis 2:1)			**Residents** ("all their host," Genesis 2:1)	
Day 1	Day and night **"God saw that the light was good."**	⟷	Sun, moon, and stars **"God saw that it was good."**		Day 4
Day 2	Heavens (between the waters)	⟷	Birds and fish **"God saw that it was good."**		Day 5
Day 3	Land and seas **"God saw that it was good."**	⟷	Animals **"God saw that it was good."**		Day 6
	"Let the earth sprout vegetation, plants yielding seed, and fruit trees bearing fruit in which is their seed, each according to its kind, on the earth." **"God saw that it was good."**		Humans (to steward/farm): "Behold, I have given you every plant yielding seed that is on the face of all the earth, and every tree with seed in its fruit. You shall have them for food." **"Behold, it was very good."**		

John Walton has shown that the creation week's primary concern is with establishing order (Walton speaks of "functions") in the world.[24] I would add that the text also points to a particular purpose instilled in the order—namely, to make the land fruitful, and in that fruitfulness to allow for all creation to feast (that is, to thrive).

The Sabbath is the crowning day of the week (Gen 2:2-3), when the week's food production can be enjoyed in rest and feasting before God. In this manner, the creation week calendar offered practical guidance for the labor and worship of the common Hebrew household. The text is not a lofty description about galaxy formation and other phenomena of little use to the average Israelite scraping together a living from the land. It is a practical calendar of weekly food production and communion with God (cf. Ex 16:22). In our society of advanced refrigeration and food storage, we no longer think about food production on a weekly basis. Jesus' prayer, "Give us this day our daily bread" (Mt 6:11), is lost on us. But for Israel, the creation week calendar—like the festival year calendar—taught a practical cadence of labor and rest in the land God had prepared for them as his stewards (Gen 1:27-28).

Conclusion

The core idea behind this reading of the text is as old as the fourth commandment: that the creation week is a festival calendar. There are several implications of this study. Let me close by highlighting two of them.

First, this reading cautions against both young-earth and old-earth efforts to read Genesis 1 as a chronology of original creation events. Where the Pentateuch does record chronology, it uses event sequencing, not timeline dating. Where the Pentateuch adds dates, those dates are added for festival alignment. Genesis 1:1–2:3 therefore shouts to us concerning Israel's cadence of labor and worship in the world God created, but the text is silent concerning the timing by which he created it. This reading leads to conclusions largely congruent with "analogical day," "literary day," or "framework" views.[25]

---

[24]John H. Walton, *The Lost World of Genesis One: Ancient Cosmology and the Origins Debate* (Downers Grove, IL: IVP Academic, 2009), 21-70.

[25]E.g., see the chapters by C. John Collins, Richard Averbeck, and Tremper Longman III in *Reading Genesis 1–2: An Evangelical Conversation,* ed. J. Daryl Charles (Peabody, MA: Hendrickson, 2014).

Second, this study shows that vocation and sabbath are the theological heart of the creation week narrative. One of the unfortunate results of the creation debates has been an overemphasis on the six days to the neglect of the seventh.[26] Reading the six days without emphasizing the seventh would be like making a blockbuster movie about the ten plagues without highlighting their culmination in the Passover. It might scratch an itch for drama, but it truncates the festival climax of the story. The Torah's calendar narratives are *festival* stories. Our primary interest should be in the festivals they inform, including the Sabbath-week calendar.

Various theological traditions differently assess the New Testament's teaching on the Sabbath, whether the Sabbath is continued, modified, or repealed in the church today. Differences over how the Sabbath develops in later history are an important further topic, but the creation week provides the anchor for that theme. The text's function as a calendar calls us to re-center our interest in the text on its practical calling of God's people to weekly vocation and communion.

---

[26]See note 3 above.

2

Galaxies, Genes, and the Glory of God

DEBORAH B. HAARSMA

A S AN ASTRONOMER, I HAVE HAD THE PRIVILEGE of studying some of the largest, most distant objects in the universe: galaxies containing trillions of stars, galaxy clusters so massive that they curve space itself, and a universe expanding at an ever-increasing rate. But there is something much closer to earth that has been on my bucket list; I've always wanted to see a total solar eclipse, and last summer I finally had my chance.

On August 21, 2017, millions of people across the country saw the eclipse. Many viewed the *partial* solar eclipse through eclipse glasses, watching the moon slowly move across the sun. But some of those millions traveled to the narrow path across the continent where they could view the *total* solar eclipse, which is another scale of drama altogether. My family planned our summer vacation to Oregon to see it. For two minutes, the whole sky went to a deep, dusky blue, while the corona streamed from the sun far brighter and larger than I expected. The moon was a black circle in the middle of the corona, looking like a hole in the sky. It was incredible! Everyone who saw the total eclipse felt a jaw-dropping sense of wonder and amazement.

Of course, those watching the eclipse understood the scientific explanation: the moon passes in front of sun. Did that scientific explanation detract from their sense of wonder? Not at all! In fact, their wonder was increased. The scientific explanation allowed them to protect their eyes and enhanced their enjoyment as they observed the astronomical bodies in motion and the blazing features of the corona.

For Christians, encounters with the natural world also have a spiritual component. Beyond a general sense of awe, we experience the glory of the Creator. We know that the universe didn't create itself, nor is it the product of some impersonal force. Rather, the heavens declare the glory of God (Ps 19:1). We know there is a person behind the universe, the same person incarnate in Jesus Christ. "In the beginning was the Word. . . . Through him all things were made" (Jn 1:1, 3). The Creator of the universe is a person who knows us and longs to be known. Our encounter with wonder is tied to the Savior we walk with every day.

Where else might our faith intersect our understanding of the natural world? How can we understand the authoritative teaching of Scripture and orthodox theology in a modern scientific culture? In this chapter I offer several reflections as a Christian and a scientist.

SCIENCE AS A CHRISTIAN VOCATION

I'm not the only scientist in my house; my husband, Loren, is also a physicist. Yes, conversations at home get nerdy sometimes (date night a few weeks ago was watching *BattleBots* on TV). As physicists, Loren and I both love how math describes the real world. It's amazing how complex calculations on a sheet of paper can correctly predict the behavior of objects, from planets to planes to positrons, with incredible precision.

This encounter with the mathematical order of the universe gives physicists a sense of wonder, not unlike the wonder that people felt watching the eclipse. Since we are Christian physicists, that wonder leads us to worship of the Creator who made the universe with incredible order. Far from the scientific explanation detracting from worship, here it prompts it. *A scientific explanation does not replace God.* Rather, it shows us more of God's works.

So it is jarring to hear some atheists shouting that science has disproved God and made religion irrelevant. Richard Dawkins and others promote a worldview of scientism, in which science is the best—or only—kind of knowledge. On March 29, 2017, atheist biologist Jerry Coyne blogged: "I don't think one can be really smart and religious at the same time. . . . Many public intellectuals—and virtually all accomplished scientists—are atheists. . . . Someone, regardless of how 'smart' they seem, is at the very least irrational if they believe in God or the attendant superstitions."[1] Coyne brashly claims that no smart, rational, or educated person would be religious! Not all atheists take such an extreme line, but enough do that the message has infiltrated our culture. The church today needs to speak against scientism and bring a thoughtful voice to the public square.

Besides, Coyne is wrong on the facts. The majority of elite scientists are *not* militant atheists but hold some form of spirituality, or at least respect religion in others.[2] The ranks of elite scientists include deeply committed Christians such as Francis Collins, who directed the Human Genome Project and now leads the National Institutes of Health. Collins shared his testimony in the bestseller *The Language of God.*[3] And he is not alone. Geneticist Praveen Sethupathy at Cornell, astronomer Jennifer Wiseman, biophysicist Ard Louis at Oxford, and biologist Jeff Hardin at Wisconsin–Madison are all great examples, among many others.[4] These scientists have impeccable research records, defying Coyne's accusation of irrationality. They are also people of deep faith; I've given names of my friends, for whom I can personally attest their commitment to Christ, their prayer life, and their involvement in local congregations (two preach occasionally). Their scientific talents clearly have not wiped out their belief in God or their obedience to Christ's teaching.

These believing scientists are not compartmentalizing their lives. They do not set aside their faith during the week, nor do they ignore science on Sunday. Rather, they see their scientific work as a natural outgrowth of their faith. How does a Christian hold to genuine faith while doing the same scientific experiments as atheist colleagues? Consider some key principles necessary to do science:

- Humans can understand nature.

- Nature operates with regular, repeatable, universal patterns.

- Experiments are needed; theories are not enough.

- Science is worth doing.

---

[1] Jerry Coyne, "Are Religious People a Bit Thick?," *Why Evolution Is True* (blog), March 29, 2017, https://whyevolutionistrue.wordpress.com/2017/03/29/are-religious-people-a-bit-thick.
[2] Elaine Howard Ecklund, *Science vs. Religion: What Scientists Really Think* (New York: Oxford University Press, 2010).
[3] Francis S. Collins, *The Language of God: A Scientist Presents Evidence for Belief* (New York: Free Press, 2006).
[4] Meet more Christians in science in R. J. Berry, ed., *True Scientists, True Faith* (Oxford: Monarch, 2014), and Ruth Bancewicz, *God in the Lab: How Science Enhances Faith* (Oxford: Monarch, 2015).

Such principles are shared by scientists of all worldviews. But each scientist has his or her own reasons for them. A Christian believes the following:

- We are made in God's image (Gen 1:27).

- Nature is not filled with capricious gods but ruled by one God in a faithful, consistent manner (Gen 1; Ps 119:89-90).

- God's creativity is free, but we are limited and fallen (Job 38).

- We are gifted and called by God to study God's handiwork (Gen 1:28; 2:19-20; Prov 25:2; Ps 19:1).

Thus, a Christian worldview naturally gives rise to the underlying principles necessary for the practice of science.[5] When I observe galaxies with a telescope, I am not setting aside my faith but using scientific methods that flow naturally from my beliefs about God. Although my methods and immediate conclusions may be the same as my atheist colleagues, I differ from them in my motivations and in the broader implications I draw from the results.

Pastors can minister to scientists in their congregations by affirming their calling and supporting their witness in a secular workplace.[6] They can also inspire students to pursue science by explaining the principles above, showing how believers can follow Christ in a scientific career.

Yet the perception of conflict persists. For many seekers, the idea that the church is anti-science is holding them back from Christ. A man emailed me at BioLogos saying, "I'm sixty-seven years old and have never believed in God. Some events in my life have caused me to look for a relationship with Jesus. The science versus faith [issue] has always been my main reason to reject God. Then I found Francis Collins online and then BioLogos. . . . The Holy Spirit is guiding me towards Christ because of BioLogos."

Pastors and theologians can remove barriers to faith by showing how the church is not anti-science. Celebrating the positive synergies between faith

---

[5]For more on worldviews and science, see chap. 2 of *Origins: Christian Perspectives on Creation, Evolution, and Intelligent Design* (Grand Rapids: Faith Alive Christian Resources, 2007), which I wrote with my husband, Loren, sharing our personal journeys and discussing the range of views Christians hold on questions of Gen 1, age, evolution, and the historical Adam. The book is useful for small groups, with discussion questions and short videos.

[6]For more on ministering to scientists, see Andy Crouch, "What I Wish My Pastor Knew About the Life of a Scientist" (Parts 1–3), BioLogos, April 29–May 1, 2013, https://biologos.org/blogs/archive/series/what-i-wish-my-pastor-knew-about-the-life-of-a-scientist.

and science will make the church more welcoming to the many unchurched people who work in science fields and already see the beauty and mathematical order in nature. Some years ago I met a biologist who shared her story with me. She grew up outside the church but loved experiencing nature, whether through walks in the forest or biology experiments. When describing how she became a Christian as an adult, she told me something I never forgot: "There was no way I would have come to faith in God if nature and science weren't a part of it." For seekers like Joyce, addressing science in church is key to their belonging and belief.

CREATION AS REVELATION ABOUT GOD

An age-old theological metaphor describes the natural world as a second "book" of God's revelation. Augustine used this metaphor, and the parallelism goes back to Psalm 19. I love how the Belgic Confession (1566) puts it: the universe is "like a beautiful book in which all creatures, great and small, are as letters to make us ponder the invisible things of God." Nature as well as Scripture is a revelation from God and about God. Scripture is our best teacher of God's character and his will for us, but the natural world displays the work of the same Creator and resonates with what we know of God from Scripture.

The universe is God's work of art, filled with beautiful things.[7] Star clusters are a great example; take a moment to search online for images of NGC 6362.[8] This cluster contains thousands of shining stars, a jewel box of brilliant colors. In such places of shining glory, the heavens truly declare the beauty of the Creator. God made over a thousand star clusters in our galaxy alone! These beautiful clusters were around long before humans could photograph and enjoy them. We get a sense of God creating for his own pleasure, in extravagant abundance.

Star clusters have more to teach us. Where do they come from? Astronomers actually observe star clusters in the act of forming from large clouds

---

[7]For more ways the universe reveals the triune God, see Deborah and Loren Haarsma, "Christ and the Cosmos: Christian Perspectives on Astronomical Discoveries," in *Christ and the Created Order*, ed. Andrew Torrance and Thomas McCall (Grand Rapids: Zondervan, 2018). See similar essays by several scientists in *Delight in Creation: Scientists Share Their Work with the Church*, ed. Deborah Haarsma and Scott Hoezee (Grand Rapids: Center for Excellence in Preaching, Calvin Theological Seminary, 2012).

[8]For example, see "Hubble Image of the Globular Star Cluster NGC 6362," European Southern Observatory, October 31, 2012, www.eso.org/public/images/eso1243d/.

of gas and dust called *nebulae* (singular *nebula*).[9] Nebulae have an artistic beauty, with swooping dust lanes and gasses glowing in many colors. Many nebulae have dark clumps of dust that act as cocoons for baby stars. Inside these clumps, gases and dust grains are swirling and slowly collapsing under gravity.[10] The hot gases exert pressure and resist the collapse, but if the clump is massive enough, gravity wins. The center of the clump collapses into a core dense enough for fusion reactions, releasing the light of a newborn star. The remaining material settles in a disk around the baby star, a disk that eventually coalesces into planets. Thus, in a star cluster we see not only artistry and abundance but also ongoing creation. God is still making new stars! The universe is continuing to develop, growing in complexity. God's creative work isn't once and done; it's an ongoing process.

Of course, God has the authority and power to say the word and a star cluster could simply pop into existence. But from what we observe here, God chooses to do something different. He uses mediated creation, working *with* creatures to create.[11] God works with materials he already made (gas and dust) and with natural processes he already used (gravity and gas pressure) to assemble a star cluster. It isn't a fast process (about one new star per year in our galaxy), and it isn't efficient (a lot of gas never gets made into stars). But in it we can see God's patience. Rather than using brute force, God seems to respect the integrity of the material he has already made. This gentle approach is not surprising given God's character of self-giving love. N. T. Wright recently observed:

> If creation comes through the kingdom bringing Jesus, we ought to expect that it would be like a seed growing secretly, that it would involve seed being sown in a prodigal fashion in which a lot went to waste apparently, but other seed producing a great crop. We ought to expect that it would be a strange, slow process which might suddenly reach some kind of harvest. We ought to expect that it would involve some kind of overcoming of chaos. Above all, we ought

---

[9]See the Hubble Space Telescope video tour of a star cluster and nebula, "Celestial Fireworks: Star Cluster Westerlund 2," HubbleSite, April 22, 2015, http://hubblesite.org/video/26/science.

[10]See an image and more information at "Astronomers Get Rare Peek at Early Stage of Star Formation," National Radio Astronomy Observatory, March 14, 2012, www.nrao.edu/pr/2012/clumpcores/.

[11]Robert C. Bishop, "Recovering the Doctrine of Creation: A Theological View of Science," BioLogos, January 1, 2011, https://biologos.org/resources/scholarly-articles/recovering-the-doctrine-of-creation-a-theological-view-of-science, and Colin Gunton, *The Triune Creator* (Grand Rapids: Eerdmans, 1998).

to expect that it would be a work of utter, self-giving love, that the power which made the world, like the power which ultimately rescued the world, would be the power not of brute force but of radical, outpoured generosity.[12]

Star clusters are just one example from nature that could inspire preaching. Pastors can lead congregations in celebrating the glory of creation and its testimony to the Creator.[13] This doesn't require scientific expertise—you can show a science video or interview a scientist from the community. At the conference leading to this volume, singer and author Andrew Peterson spoke with passion of encountering creation in the beehives in his own backyard. Whether through preaching, worship, or education, there are many positive ways to engage science in the life of your congregation.[14]

SCRIPTURE AND CREATION

I mentioned above that the formation of a star cluster was slow, but I didn't say *how* slow. A single cluster takes a few million years to form. Some star clusters formed billions of years ago. The ages of many astronomical objects—from the earth to the universe as a whole—have been measured, and all point to billions of years.[15] God's creation is ancient. But how can that fit with the six days of Genesis 1? This was a challenging question for me. I grew up in a Christian home and a wonderful church. We believed the earth was young, because that was the only Christian view we knew. As a graduate student I wanted to study astronomy, yet I loved the Bible and didn't want to ignore Genesis. How do we reconcile the clear testimony of God's creation with the authoritative and inspired word of God in Scripture?

A helpful starting point here is the metaphor of the "two books" of revelation. Since God reveals himself in both nature and Scripture, these cannot be in

---

[12]N. T. Wright, "If Creation Is Through Christ, Evolution Is What You Would Expect," BioLogos, April 25, 2017, video, 4:01, https://biologos.org/blogs/guest/nt-wright-if-creation-is-through -christ-evolution-is-what-you-would-expect.

[13]For more on preaching, see pastor Scott Hoezee, *Proclaim the Wonder: Engaging Science on Sunday* (Grand Rapids: Baker Books, 2003), and pastor John Van Sloten, *Every Job a Parable: What Walmart Greeters, Nurses, and Astronauts Tell Us About God* (Colorado Springs: NavPress, 2017).

[14]Deborah Haarsma, "Engaging Science in the Life of Your Congregation," Fuller Studio, https:// fullerstudio.fuller.edu/engaging-science-life-congregation/. Originally published as "Thinking Science and Faith Together," *Theology, News & Notes* (Fuller Seminary), Spring 2013.

[15]For a short video and links to articles, see "How Old Is the Earth?," BioLogos, September 11, 2014, video, 3:12, https://biologos.org/resources/audio-visual/how-old-is-the-earth.

conflict; both must speak truly of God and his creation. Yet both are *interpreted*. Science is our human interpretation of nature, and we don't always get it right. And the church doesn't always agree on biblical interpretation. If one or both are mistaken in interpretation, then there will be conflict at the human level. The right approach to conflicts, then, is not simply to reject Scripture on the basis of one scientific finding, nor to reject nature on the basis of one interpretation of Scripture. Rather, we must dig deeper into the interpretation of each.

I experienced a turning point when I discovered the work of Old Testament scholars and theologians such as John Stek and, later, Tremper Longman, Richard Middleton, and others. I learned that Genesis was written in a prescientific age and that the ancient cultures of Egypt and Babylon had a very different view of the natural world than we do. They believed the earth was flat and pictured the sky as a solid dome with water above for rain. I had always been puzzled by day two of creation, but now finally understood that Genesis 1:6-8 reflects this ancient understanding of a sky-dome firmament. John Walton shows how the seven-day pattern in Genesis 1 fits the temple narratives of ancient cultures.[16] Thus, God revealed this important text within a particular cultural context so that the Hebrews could understand it. John Calvin notes, "For who even of slight intelligence does not understand that, as nurses commonly do with infants, God is wont in a measure to 'lisp' in speaking to us? Thus such forms of speaking do not so much express clearly what God is like as accommodate the knowledge of him to our slight capacity."[17]

While accommodated to a prescientific picture of the cosmos, Genesis 1 is very different religiously from the views of the surrounding cultures. They pictured the world arising from a battle between the gods, with humans as an afterthought. In its theology, the text is clearly not accommodating to the times but describing a good world made by one sovereign Creator. This gives good reason to see the theological message as the primary intent for the original Hebrew author and audience, and it can be so for us as well. The main point is not the days and the physical structures but fundamental theological truths: that one God is the Creator, that

---

[16]John H. Walton, *The Lost World of Genesis One: Ancient Cosmology and the Origins Debate* (Downers Grove, IL: IVP Academic, 2009).

[17]John Calvin, *Institutes of the Christian Religion*, ed. John T. McNeill, trans. Ford Lewis Battles (Philadelphia: Westminster, 1960), 1.13.1.

created things are not gods in themselves, and that humans have a special place and calling within God's creation.

Of course, Christians disagree on how to interpret Genesis 1 and have proposed many other views. I offer this interpretation as one way to uphold the authority and inspiration of Scripture while affirming the evidence in God's creation.

For pastors, there are risks in addressing a controversial topic like origins. But there are also risks if you don't. Young people are watching, and they need to hear how this passage can be true and relevant in today's scientific world. Their university science professor might tell them that science is in direct conflict with the Bible, and if they hear the same at church, they will be forced into a false choice between faith and science. A young man named Connor wrote to BioLogos when he was sixteen years old:

> I have always attended Christian schools, where I was taught to affirm a young-earth creationist view. . . . I decided to research the Big Bang a bit more, and I was overwhelmed by the strength of the evidence. . . . I became convinced that the Bible was not completely true. . . .
>
> Then I found BioLogos. They addressed all of my questions, showing a respect for the Bible's authority and the findings of science—even related to evolution. I was so excited that I didn't have to choose between science and God. . . . The more I learn about evolution, the more I just want to praise God for his magnificent creation.[18]

A good discussion of Scripture and science can *deepen* the faith of young people like Connor. They need adults to come alongside them in their questions.

THE END OF THE UNIVERSE

What is the end of God's good world? This is a biblical and theological question and also a physical question about the universe. Astronomers predict the following:

- In four or five billion years, our galaxy, the Milky Way, will collide and merge with the nearby Andromeda Galaxy.

- In about eight billion years, the sun will have ballooned out to a red giant star large enough to envelop and destroy the earth.

---

[18]Connor Mooneyhan, "How Science Shook My Faith," BioLogos, April 7, 2015, https://biologos .org/blogs/archive/how-science-shook-my-faith.

- In a few hundred billion years, all nearby galaxies will have collided and merged together.

- Ultimately, the universe will continue to expand, with stars dying until the universe is cold and dark (the "Big Freeze"). The universe may accelerate in its expansion to the point that all objects are torn apart by space itself (the "Big Rip").

However, these predictions are based on two big assumptions, that (1) our current understanding of the relevant physical laws is correct and complete, and (2) those laws will continue to act in the same way for an indefinite time into the future. Certainly the first assumption is debatable. Over trillions of years and more, physical processes that have a minor impact today may grow to have a major impact that we cannot predict. It is presumptuous to think our current understanding of physics is complete. As for the second assumption, science alone can't guarantee that the natural processes we see working today will continue to work tomorrow.

As Christians, we have clear teaching in Scripture against the second assumption. We believe that God will change things up in the future and bring about a new heaven and a new earth. The description of the new creation in Scripture is highly symbolic, but it paints a picture sufficiently different from today that new laws of physics will likely be at work. Christ will come again and make all things new. We need not fear that the universe will end in cold, dark emptiness, but can look forward in hope to a renewed or re-created world where we will live in the direct presence of God.

The Goodness of Creation

What does God mean in calling creation "good" in Genesis 1? I've learned from theologians and biblical scholars that this word means "fit to God's purpose," obedient to the Creator and supporting and fulfilling God's intentions as he crafts the world.[19] God's purposes may not be ours, and thus we must take care in what we assume "good" to mean. It may not fit our human

---

[19]Richard Middleton, *The Liberating Image: The Imago Dei in Genesis 1* (Grand Rapids: Brazos, 2005), 77, 266; John H. Walton, *The Lost World of Adam and Eve* (Downers Grove, IL: IVP Academic, 2015), 169-70; Iain Provan, *Seriously Dangerous Religion: What the Old Testament Really Says and Why It Matters* (Waco, TX: Baylor University Press, 2014), 283.

ideas of "perfect." Here are three case studies from astronomy and physics on what goodness in creation looks like.

Gravity. The strength of the force of gravity is a fundamental property of the universe. It appears to be fine-tuned to a precise value to allow life to exist. Here's the argument:

1. Consider even simple life, like bacteria. To survive, such life requires

    a. a variety of atoms (hydrogen, carbon, oxygen, etc.) and

    b. a stable energy source (a long-lived star like our sun).

2. Stars are the source of atoms (carbon and oxygen forms there) *and* of stable energy.

3. Thus, the existence of life requires long-lived stars.

4. Stars form from clumps of gas and dust, but

    a. the strength of gravity cannot be too small (clumps would never collapse) and

    b. the strength of gravity cannot be too large (clumps would build into large stars that burn too quickly).

5. Thus, the strength of gravity needs to be just right (fine-tuned) in order to make a good, fitting home for us.

The strength of gravity is just one example of fine-tuning; many other parameters of the universe are also set just right for life. As seen through the eyes of faith, the universe is good in that it is crafted and designed for life, fulfilling God's intention of creating a fitting home for us.[20]

Gravity is important not only for the formation of stars but also for the assembly of galaxies containing billions of stars. Astronomers today can simulate how these galaxies assemble over very long timescales via natural processes. This is evidence that God used natural processes rather than miracles to make galaxies—but he is no less the Creator. He crafted and designed a natural system including atoms and gravity that over time produces galaxies in a wealth of complexity, abundance, and variety. The universe is designed to assemble complexity by natural processes. Charles Kingsley, a

---

[20]I give this fine-tuning argument not as a logical proof for God but rather as a "design discourse" in which an observation leads to a perception of design. See Alvin Plantinga, *Where the Conflict Really Lies: Science, Religion, and Naturalism* (New York: Oxford University Press, 2011), chap. 8.

nineteenth-century Anglican priest, wrote, "We knew of old that God was so wise that He could make all things; but behold, He is so much wiser than even that, that He can make all things make themselves."[21]

Supernovae. Consider the explosion of a star at the end of its life, a *supernova* (plural *supernovae*). For a while, the explosion is brighter than a billion stars! Then the heat fades, the debris slowly expands, and thousands of years later we see the remnants of the event.[22] Clearly stars do not survive forever. Rather, they have a cycle of "life" and "death" analogous to creatures on earth, with the remnants of one generation laying the seeds for the next. God appears to have designed the natural laws so that objects in this universe, whether stars or seagulls, do not last forever.

A supernova is a violent explosion. If one happened in a nearby star, life on earth would be in danger. Dangers on earth, such as earthquakes and violent storms, are sometimes called "natural evil" or are said to be a result of human sin. However, such dangerous situations were around long before humans; they appear to be part of God's good creation from the beginning.[23] In Scripture, God claims credit for dangerous storms and earthquakes. Apparently, *good* does not mean "safe" or "tame" but can include things that are wild and dangerous. The wild places of this earth, such as Mount Everest and Death Valley, are God's good creation, but they are dangerous and best approached with caution. Although supernovae are dangerous, they disperse the carbon and oxygen that formed in stellar cores into nebulae, where the elements become parts of planets and ultimately ingredients for our bodies. God made wild and dangerous things that serve his purposes, even if they are not safe.

Snowflakes. Consider the six-sided snowflake. The orderly symmetry of the snowflake is beautiful, but what people love even more is that each snowflake is unique. That uniqueness comes not from predictable order but from randomness. Here *random* does not mean "purposeless" or "meaningless," but simply means "scientifically unpredictable." As each flake falls through the air, the winds bounce it around in an unpredictable way so that the crystal grows slightly differently and acquires a unique shape. Thus the beauty of

---

[21]Charles Kingsley, "The Natural Theology of the Future," read at Sion College, January 10, 1871, www.online-literature.com/charles-kingsley/scientific/7/.

[22]For an animated video, see "Crab Supernova Explosion," Hubble Space Telescope, December 1, 2005, www.spacetelescope.org/videos/heic0515a/.

[23]Justo L. González, *Creation: The Apple of God's Eye* (Nashville: Abingdon, 2015), 57.

snowflakes is due to a combination of order and randomness. God seems to have incorporated randomness into the system to accomplish his purpose in creating beautiful, unique snowflakes. This purposeful randomness brings about the beautiful variety he intends.

To summarize, I've given examples from physics and astronomy that display God's good creation as follows:

- glorious

- mathematically ordered

- beautiful

- extravagantly abundant

- ongoing, with cycles of life and death

- mediated rather than efficient

- ancient

- fine-tuned for life

- designed to assemble complexity

- wild and dangerous rather than tame

- purposefully random rather than fully predictable

SCIENCE OF EVOLUTION

There are many important issues at the intersection of science and faith today, from climate change to genetic engineering. One issue high on the minds of many evangelicals, and of interest to me as an astronomer, is origins. In particular, how do we as Christians engage the issue of evolution?[24]

The word *evolution* has acquired multiple meanings. Some loud atheist voices today use the word *evolution* to describe an atheistic worldview. Philosophers Alex Rosenberg and Tamler Sommers write, "Darwinism thus puts the capstone on . . . denying that there is any meaning or purpose to the universe, its contents and its cosmic history."[25] All Christians would

---

[24]See Haarsma and Haarsma, *Origins*; and my chapter in *Four Views on Creation, Evolution, and Intelligent Design*, ed. J. B. Stump (Grand Rapids: Zondervan, 2017).
[25]Tamler Sommers and Alex Rosenberg, "Darwin's Nihilistic Idea: Evolution and the Meaninglessness of Life," *Biology and Philosophy* 18, no. 5 (2003): 653.

oppose such nihilism. But this is not a scientific statement at all. It is a worldview claim that goes far beyond the science itself.

Scientists use the word *evolution* to refer to a scientific process, much as they use terms like *photosynthesis* or *plate tectonics*. The science of evolution describes how all species are related to each other and diversified over time. Both Tim Keller and C. S. Lewis emphasize the importance of disentangling evolution as a scientific process from evolution as an atheistic worldview.[26]

A third use of the term *evolution* places this scientific process in a Christian context, in a view called *evolutionary creation*. In this view, the natural process of evolution is crafted and governed by God to create the diversity and interrelation of all life on earth. At BioLogos, we present evolutionary creation as a faithful option for Christians to consider and discuss alongside other views.

Christians today discuss many views on origins, which fall into a few broad types, as seen in table 4.

Table 4. Types of views on origins

	Atheistic evolution	Youth-earth creation	Old-earth creation	Evolutionary creation
God created earth and all life	No	Yes	Yes	Yes
Earth is billions of years old	Yes	No	Yes	Yes
Evolution describes how life developed	Yes	No	No	Yes

The three Christian views agree on the most important thing: the God of the Bible is the Creator of all. But Christians disagree on *how* God went about doing it. Evolutionary creationists argue that there is strong evidence for the great age of the earth and for evolution as a scientific process, and that this picture is consonant with God as Creator, with faithful readings of Scripture, and with orthodox theology.

---

[26]Tim Keller, "Creation, Evolution, and Christian Lay People," BioLogos, 2009, https://biologos .org/resources/scholarly-articles/creation-evolution-and-christian-laypeople; C. S. Lewis, Letter to Captain Bernard Acworth, September 23, 1944, quoted in Gary B. Ferngren and Ronald L. Numbers, "C. S. Lewis on Creation and Evolution," *Perspectives on Science and Christian Faith* 48 (March 1996): 28-43, www.asa3.org/ASA/PSCF/1996/PSCF3-96Ferngren.html.

In the natural world, God has left abundant information about how species formed, showing that all species are related to each other and modified and diversified over time. Charles Darwin developed the idea in the mid-1800s from evidence available then. Scientists at that time knew of some fossils of ancient creatures, as well as the clustering of species in certain places on earth ("biogeography"—think of the finches Darwin studied on the Galapagos Islands) and the similarities of features in different species ("comparative anatomy"—think of the similar body plans of humans and chimpanzees). From this evidence, Darwin proposed that all species were descended from a common ancestor, and he figured out a key mechanism by which they evolved: natural selection.

In the century and a half since Darwin, tremendous discoveries have been made. Not only have scientists found more mechanisms and more evidence, but they have found entirely new lines of evidence that confirm Darwin's predictions and improve on his model. Consider whales as an example. On the basis of comparative anatomy, Darwin predicted that whales were descended from land mammals. For a century there was little fossil evidence to support this, but in the past few decades many new fossils have been discovered of early whales, clearly showing the transition from land mammals to ocean mammals over millions of years. And an entirely new line of evidence—genetics—confirms this. Darwin had no knowledge of DNA, and yet the genomes of many species today show the close relationship of whales to land mammals.[27] Whales are not a distant relation to mammals like the platypus is; rather, they are closely connected. In fact, a cow is more closely related to a whale than to a horse!

These multiple lines of scientific evidence lead the vast majority of biologists to accept evolution as a scientific process by which new life forms developed.[28] Many Christians now accept this evidence and affirm evolutionary creation, that God brought about the diversity of life on earth through this natural process.[29]

---

[27]For example, see fig. 1 in Andrew D. Foote et al., "Convergent Evolution of the Genomes of Marine Animals," *Nature Genetics* 47 (2015): 272-75, www.nature.com/ng/journal/v47/n3/full/ng.3198.html.

[28]Lee Rainie and Cary Funk, "An Elaboration of AAAS Scientists' Views," Pew Research Center, July 23, 2015, www.pewinternet.org/2015/07/23/an-elaboration-of-aaas-scientists-views/.

[29]For more on the scientific evidence for evolution from Christian biologists, see a short video and links to articles at "How Evolution Works, Part 1," BioLogos, https://biologos.org/resources/audio -visual/how-evolution-works-part-1, including a link to Dennis Venema's blog series "Evolution Basics" at BioLogos.org. See also Darrel R. Falk, *Coming to Peace with Science* (Downers Grove, IL:

Yet many Christians disagree with evolutionary creation, often because of biblical and theological concerns. And that's to be expected, since origins is not a gospel issue. The belief that God is the Creator is essential to the Christian faith,[30] but the timing and methods God used are not essential and Christians will honestly disagree.

How we talk about our differences on origins matters—for the church and the world. The speakers and respondents at the conference from which this book is derived embodied constructive and gracious dialogue within the church. Pastors can bring this type of dialogue to local churches, building a culture in which Christians can discuss differences on origins charitably while affirming shared belief in the Creator and the authority of Scripture. We need to do the hard work of following Christ's call to unity within the church.[31] A culture of acrimony and mudslinging in the church is a poor witness to the watching world. At BioLogos, we engaged in a multiyear dialogue with another Christian organization that holds the old-earth creation view, Reasons to Believe. We worked to dialogue and worship together in spite of our differences; it was a privilege to experience Christian unity in the midst of firm disagreements about significant questions. Our dialogues are now published and provide a useful introduction to old-earth and evolutionary creation views for pastors and theologians.[32]

Young people are watching. I appeal to Christian leaders to be honest with students about the strength of the evidence for evolution as a scientific process, and to teach them to distinguish evolution as science from evolution as an atheistic worldview. Even if you hold one of the other views, by presenting young people with the range of faithful Christian options, you will show them they do not have to choose between their faith and scientific evidence.

InterVarsity Press, 2004); Denis Lamoureux, *Evolutionary Creation* (Eugene, OR: Wipf & Stock, 2008); and Denis R. Alexander, *Creation or Evolution: Do We Have to Choose?* (Oxford: Monarch, 2014).

[30]Deborah Haarsma "Essentials of Creation," BioLogos, September 12, 2017, https://biologos.org/blogs/deborah-haarsma-the-presidents-notebook/essentials-of-creation-a-response-to-the-gospel-coalition.

[31]See the blog series by theologian Ross Hastings, "Ephesians 4:1-6: A Call of Christian Unity," BioLogos, February 7–11, 2011, https://biologos.org/blogs/archive/series/ephesians-41-6-a-call-of-christian-unity.

[32]Ken Keathley, J. B. Stump, and Joe Aguirre, eds., *Old-Earth or Evolutionary Creation? Discussing Origins with Reasons to Believe and BioLogos* (Downers Grove, IL: IVP Academic, 2017).

THEOLOGY AND EVOLUTION

What implications does evolutionary creation have for orthodox Christian theology? The insights from physics and astronomy above have some parallels here.

Divine action in evolution. All Christians affirm that God governs natural processes *and* performs miracles. But which path did God use when creating new species? In young-earth and old-earth creationist views, God uses miracles to create many species, but in evolutionary creation God works through natural processes to create every new species. Do these natural mechanisms mean that God is more distant? No. In other areas of nature, such as the mathematically ordered law of gravity that keeps Earth in orbit around the sun, all Christians celebrate natural process as God's creation. Evolutionary creationists see the process of evolution in the same way. *A scientific explanation does not replace God.* In fact, we see mediated creation in life on earth as we saw it in the universe. Just as God uses existing materials and slow natural processes to create new stars, so God can work over time with the DNA of existing life forms and environmental conditions to create new species.

People sometimes ask whether natural processes, without additional miracles, are able to increase the information in DNA and the complexity of new life forms. Evolutionary creationists answer yes. Just as we celebrate the amazing sufficiency of the fine-tuned parameters of the universe for the natural assembly of complex new galaxies, so we can celebrate the amazing sufficiency of biochemical processes for the natural assembly of complex new species.

A final theological concern relates to the random processes central to evolutionary development, particularly genetic mutation. How could randomness be part of God's orderly activity? As we saw above, the problem comes from confusing our colloquial use of *random* to mean "purposeless" with the scientific use of *random* to mean "unpredictable." Scientific randomness is essential for a host of biological processes, from building immunity to assembling cells.[33] Just as we saw in snowflakes, randomness is key for bringing about abundant variety. Evolution produces not just a few kinds of flowers but an extravagant abundance of flowers with every

---

[33]Kathryn Applegate, "Understanding Randomness," BioLogos, 2013, https://biologos.org/blogs/kathryn-applegate-endless-forms-most-beautiful/series/understanding-randomness.

variation of size, shape, color, and scent.[34] We can praise God not only for the beauty of individual flowers but for the beauty of the system God uses to create all flowers.[35]

Goodness and natural evil in evolution. I have highlighted some of the beautiful things in creation but also noted that God's good creation includes things that don't fit our ideas of "perfect" or "beautiful." Some creatures are just ugly or awkward, such as the ostrich. Yet God celebrates ostriches as his creation even in their silliness (Job 39:13-18).

A deeper concern is suffering and death—why would a loving, powerful God allow it? This huge question troubles all of us when we watch people suffer. The question has been asked for millennia, long before evolutionary science, so I don't claim to provide a full answer here. Yet there is a new aspect of the question in that evolutionary creation involves animal suffering, death, and extinction from the beginning as part of the system. We found above that wild and dangerous things, such as supernovae and earthquakes, can be part of God's good creation. Animal predation could be the same way. God celebrates the predatory hawk whose young feast on blood (Job 39:26-30) and takes credit for feeding the lions (Ps 104:19-23). The Bible does not teach that animals were originally immortal or that animal death is a result of sin. Thus, predation may be part of God's original creation.[36] Just as stars are being continually created and do not last forever, so God's plan from the beginning could have been to have species rise, live, and then go extinct to make room for new species. Ultimately, God's new creation will surpass the first and be free of animal predation (Is 11:6-9).

ADAM AND EVE

The questions around human origins are some of the most challenging questions raised in evolutionary creation because they are tied to the interpretation of key biblical texts and to key doctrines of the image of God, sin, and

---

[34]Annie Dillard celebrates biological abundance in *Pilgrim at Tinker Creek* (New York: Harper's Magazine Press, 1974).

[35]Deborah Haarsma, "Praising God for His Work in Evolution," in *How I Changed My Mind About Evolution: Evangelicals Reflect on Faith and Science*, ed. Kathryn Applegate and J. B. Stump (Downers Grove, IL: IVP Academic, 2016).

[36]For more on animal suffering, see Michael Murray, *Nature Red in Tooth and Claw: Theism and the Problem of Animal Suffering* (New York: Oxford University Press, 2011).

salvation. The past few years have seen several new books on the question of Adam, providing many resources for pastors and theologians.[37]

Evolutionary creationists are drawing attention to the evidence in God's creation that humans evolved. Genetics and fossil evidence clearly point to humans arising through an evolutionary process, with modern *Homo sapiens* emerging roughly two hundred thousand years ago as a population of several thousand individuals. These findings are quite different from the traditional view of Adam and Eve as sole progenitors living about ten thousand years ago.

The new scientific findings have prompted multiple proposals for understanding Adam and original sin, including proposals in which Adam and Eve are not historical people. Yet the scientific evidence still works with historical models. For example, Adam and Eve could have been two real historical people as leaders of the first group of humans; in this view, they would have been the first to sin, and their descendants would have interbred with other early humans so that Adam and Eve are ancestors of us all.

I'm encouraged to see several models for Adam and Eve that uphold orthodox theology and the authority of Scripture while also affirming the evidence in the created order. In all of these models,

- Humans have sinned and need a savior, regardless of when the first sin happened.

- Humans are made in God's image, even though God made our bodies through natural mechanisms rather than a miracle.

- Humans are set apart from the animal kingdom, having clearly different capacities, even though we share common ancestry with other life.

- Humans formed all together. Modern genetics shows that all races and ethnicities are one human family.

The discussion of Adam and Eve is going strong and will continue, as pastors, theologians, and churches wrestle with the implications of scientific

---

[37]A good book to start with is *Four Views on the Historical Adam*, ed. Matthew Barrett and Ardel B. Caneday (Grand Rapids: Zondervan, 2013). For an overview of several views within evolutionary creation and links to many articles and books, see "Were Adam and Eve Historical Figures?," BioLogos, https://biologos.org/common-questions/human-origins/were-adam-and-eve-historical-figures.

discoveries. It is a challenging discussion, but one with many hopeful avenues for discovering how God created us.

Human Significance

Finally, what does it mean to be human in light of modern science? Whether or not humans share ancestry with animals, astronomers have found that humans are a small piece of an incredibly vast cosmos.

Think back to a time when you saw a dark, clear night, with a sprinkling of stars on a black background. It turns out that blackness isn't empty. When powerful telescopes zoom in on a tiny piece of it and gather the faintest light, they discover it is filled with thousands of galaxies.[38] The galaxies are large and small, elliptical and spiral, red and blue, near and far. Every patch of night sky reveals the same picture. Our own Milky Way is one of billions of galaxies in the universe, and we are but a tiny part of the Milky Way.

Does our tiny size mean we are insignificant? Carl Sagan of the old *Cosmos* series was a great astronomer and teacher, but his atheism became nihilistic on this point. He wrote, "Our planet is a lonely speck in the great enveloping cosmic dark. In our obscurity, in all this vastness, there is no hint that help will come from elsewhere to save us from ourselves."[39] Sagan viewed the universe through an atheistic lens and drew the conclusion that we were insignificant and forgotten. When considering the same universe through a Christian lens, the picture is completely different. In Psalm 103:11-12 we read:

> For as high as the heavens are above the earth,
>   so great is [God's] love for those who fear him;
> as far as the east is from the west,
>   so far has he removed our transgressions from us.

The passage doesn't say, "As high as the heavens are, so insignificant are humans." No, it instead points to the vast universe as a metaphor for the vastness of God's grace. His forgiveness is so huge that we can picture him removing our sins to the ends of the universe.

---

[38]"Hubble Goes to the eXtreme to Assemble Farthest-Ever View of the Universe," Hubble Space Telescope, September 25, 2012, www.nasa.gov/mission_pages/hubble/science/xdf.html.
[39]Carl Sagan, *Cosmos* (New York: Random House, 1980), 193.

Ultimately, the essence of human significance is in Jesus Christ. He is the divine Word, the sovereign Creator of all (Jn 1:1-3). Yet he took on a human body and "moved into the neighborhood" (as translated in *The Message*). "The Word became flesh and made his dwelling among us. We have seen his glory" (Jn 1:14 NIV). It is for humans that Christ came to earth, and for humans that he died and rose again. God's immense self-sacrificing love for each of us is the truest display of his glory. And that is why I fall in love with Jesus Christ over and over again.

3

Mere Creation

*Ten Theses (Most) Evangelicals
Can (Mostly) Agree On*

TODD WILSON

NOT LONG AGO A PASTOR FRIEND CALLED, asking for help. "I'm preaching through Genesis 1–11," he said, "and I need some advice on the whole creation and evolution thing." There was anxiety in his voice. He wasn't sure how preaching on origins was going to go in his church setting— or whether he would even survive! Understandably so. There is hardly a more controversial subject among evangelical Christians.

Several years earlier a rumor circulated within my congregation along the following lines: "Pastor Todd thinks we came from apes!" My congregation was, historically speaking, on the conservative side of many theological issues, this one included. In its not-too-distant past the church had embraced six-day, young-earth creationism as its (unofficial) teaching position. Needless to say, when they came to terms with the fact that their relatively new and fairly young senior pastor held to a version of evolutionary creation, there was a bit of congregational heartburn.

This tension-filled season in the life of our church provided the leadership with a good occasion to engage in serious conversations about origin issues. As part of this process, I articulated my own views in the form of a seventeen-page white paper we discussed as a leadership. More importantly, we grappled with our doctrinal boundaries as a local church, asking the question, What degree of diversity will we allow on this issue? And equally as important, given our diversity, what can we still affirm together as a unifying doctrinal core?

Those were good conversations. The upshot was the development of a series of ten theses on creation and evolution that we believe (most) evangelicals can (mostly) affirm. We weren't looking for perfect unanimity. We knew we weren't going to find that. Our aims were more modest. We were looking for 100 percent of our leadership team to feel at least 80 percent good about all ten theses. This freed everyone from the burden of absolute agreement with everything said; it also allowed for diversity of opinion and even dissent on various points while still maintaining a unified position.

Our ultimate goal with these ten theses was to maintain the "unity of the Spirit through the bond of peace" (Eph 4:3) and to prioritize the gospel as of "first importance" (1 Cor 15:3). It was important for us to arrive at a position on creation and evolution that was in keeping with that faithful Christian saying, *In essentials, unity; in nonessentials, liberty; in all things, charity.*

In this essay I share with you our ten theses on creation and evolution—or what we call Mere Creation. This is not what young-earth creationists believe, or old-earth creationists believe, or advocates of intelligent design believe, or evolutionary creationists or theistic evolutionists believe, but what most (evangelical) Christians, at most times, have believed and should believe about creation. I provide brief commentary on each thesis to give some context and rationale.

Ten Theses

1. The doctrine of creation is central to the Christian faith.

Historically speaking, evangelicals have struggled to take the doctrine of creation seriously. Our love has been soteriology and Christology, not creation. I won't trace the reasons for this doctrinal penchant here. Suffice it to say that evangelicals have not engaged with the doctrine of creation robustly. It has been a side issue.

But our neglect of the doctrine of creation is not only because our attention has been elsewhere; we have sometimes downplayed the doctrine of creation for the sake of ecclesial cohesion. We've categorized the doctrine as a "secondary" or "tertiary" issue in an attempt to preserve church unity. Why break fellowship over an issue not directly related to the mission of the church or the salvation of souls?

In many ways, this approach has served evangelicalism well. One of the strengths of evangelicalism is its ability to forge common cause out of theological diversity. And yet the danger is that our toleration for doctrinal differences becomes an indifference to doctrine.[1] I fear this is precisely what has happened to the doctrine of creation within evangelicalism—a "don't ask, don't tell" policy.

Of course, not every doctrine is central to the Christian faith. Some are nearer to the core or closer to the periphery than others. Angelology isn't central. Nor are certain aspects of eschatology. But the doctrine of salvation is; so too the doctrine of God, the doctrine of the Spirit, and the doctrine of Christ.

We should add to this list the doctrine of creation for the simple reason that it addresses some of the fundamentals of our faith—the reason for and nature of the world God has made, as well as the reason for and nature of the creatures God has made, not least those creatures made in God's image. To reflect deeply on these questions is not wasted time, because the doctrine of creation is central to the Christian faith.

2. The Bible, both Old and New Testaments, is the Word of God, inspired, authoritative, and without error. Therefore whatever Scripture teaches is to be believed as God's instruction, without denying that the human authors of Scripture communicate using the cultural conventions of their time.

This second thesis is intended to be a straightforward affirmation of what many would still consider to be a classic evangelical view of Scripture. It is included as one of the ten theses because I have found it helpful in origin discussions to begin with a full-throated affirmation of the inspiration, authority, and inerrancy of the Bible. This is especially true for those who are sympathetic to evolutionary creation since they are sometimes unfairly portrayed as sitting loosely to Scripture.

I've also found that Christians who reject an evolutionary account of origins do so not primarily because they find the science unconvincing but because they have come to the conclusion that such a view will inevitably undermine the authority of the Bible. The fear is that embracing evolution leads to compromising biblical authority. That is why it is helpful to begin origin conversations with a robust affirmation of the doctrine of Scripture.

---

[1] I owe this way of putting things to a comment made by David Wells, *Courage to Be Protestant* (Grand Rapids: Eerdmans, 2008), 8.

The thrust of this thesis is that whatever the Bible teaches, God teaches, or whatever Scripture *asserts* (as distinct from what Scripture merely affirms) is to be believed as what God intends to say. It's not a viable option for those committed to the authority of Scripture to say, "I know the Bible teaches this, but I don't believe it." Or, "I know God's Word says it, but I don't buy it." These aren't viable options, at least not for self-identifying evangelicals.

In saying this, however, we want to avoid implying that God did an "end run" around the authors of Scripture. No amount of stress on a "high" view of the Bible should cause us to inadvertently downplay the human side of the equation. As Don Carson nicely puts it, "The Bible is an astonishingly human document."[2] We also do not want to suggest that a robust view of Scripture leaves no room for the authors to communicate divine truths through the cultural conventions of their time.

The Bible is indeed a "God-breathed" book (2 Tim 3:16). Yet commitment to biblical inspiration, and even inerrancy, should not undercut the human dimension of Scripture. A classic evangelical doctrine of Scripture allows for both. Yes, the authors of Scripture were "carried along by the Holy Spirit," but they "spoke from God" from within their own historical particularity and cultural conditioning (2 Pet 1:21).

When we read the Bible, then, not least when we read the creation accounts in Genesis 1–2, we want to know the author's intention as expressed in the text written, even if this doesn't exhaust a faithful handling of Scripture. At root, we want to know what this particular author meant to say, at this particular time, with these particular cultural conventions.

3. Genesis 1–2 is historical in nature, rich in literary artistry, and theological in purpose. These chapters should be read with the intent of discerning what God says through what the human author has said.

From thesis two about the authority of Scripture, we move to thesis three about the teaching of Scripture—from what Scripture is to what Scripture says. This is where all the proverbial bugs come out of the rug because here we move toward interpretation and must take into account the literary genre of these opening chapters of the Bible. For, as Old Testament scholar

---

[2]D. A. Carson, "Approaching the Bible," in *Collected Writings on Scripture* (Wheaton, IL: Crossway, 2010), 23.

Ronald L. Giese Jr. has said, "Accurate interpretation hinges then on recognizing the genres used in the Bible to communicate God's revelation."[3]

Of course, there is much debate about how to interpret Genesis 1–2, and the question of the literary genre of these chapters is something of a watershed. All too often the question is posed as an either-or: Is Genesis fact or fiction? Is it historical or theological? Does it reveal literary crafting or is it describing actual historical events?[4]

What is urged by this third thesis is the need to have a *balanced* approach to the question of the literary genre of Genesis 1–2. This means allowing for the fact that the text is a carefully crafted composite genre with all three elements—literary, historical, theological—present. Incidentally, could it be that the composite nature of Genesis 1–2 explains why among patristic witnesses "we do not find a univocal reading or a single method" for interpreting these early chapters of the Bible?[5]

It would be hard to expunge *all* historical referents from Genesis 1–2. Clearly, the text is intended to be read as an historical account, at least at some level. This isn't ancient mythology or folklore. More is going on. And yet a close reading of these texts reveals rich literary artistry. This isn't the kind of "just the facts" reporting you find at CNN or in the *New York Times*. These texts reveal sophisticated literary design in a myriad of ways, perhaps most obviously in the structuring of the seven days of creation.[6]

Yet it seems clear that the author's aim is ultimately *theological*—to say something about God, the nature of the world, and the identity and destiny of human beings who are created in his image (Gen 1:27). Whatever else is going on in these chapters, they are teaching us something about God as Creator and the world he has created. The point is not ultimately about supernovas or greenhouse gases or horticulture but about "God the Father Almighty, Maker of Heaven and Earth," as the Apostle's Creed puts it.

---

[3]Ronald L. Giese Jr., "Literary Forms of the Old Testament," in *Cracking Old Testament Codes*, ed. D. Brent Sandy and Ronald L. Giese Jr. (Nashville: Broadman & Holman, 1995); cited in Richard F. Carlson and Tremper Longman III, *Science, Creation and the Bible: Reconciling Rival Theories of Origins* (Downers Grove, IL: InterVarsity Press, 2010), 56.

[4]So a recent collection of essays debating the literary genre of Gen 1–11 is given the title *Genesis: History, Fiction, or Neither?* (Grand Rapids: Zondervan). In fairness, the essays are much more nuanced than the either-or title might suggest.

[5]The comment comes from Peter C. Bouteneff, *Beginnings: Ancient Christian Readings of the Biblical Creation Narratives* (Grand Rapids: Baker Academic, 2008), ix.

[6]Cf. Henri Blocher, *In the Beginning: The Opening Chapters of Genesis* (Downer Grove, IL: InterVarsity Press, 1984), 39-59.

Of course, affirming that Genesis 1–2 is a composite genre—with both historical and literary elements and a clear theological purpose—doesn't immediately solve issues of interpretation. Scholars will undoubtedly continue to debate the meaning of these chapters. Some will insist that they are more literary than historical, others that they are more theological than historical, and still others that they are more historical than literary. But as we seek common ground as Christians, we should at least begin with a shared commitment to authorial intention and agreement that the genre of Genesis 1–2 is complex and arguably composite.

4. God creates and sustains everything. This means that he is as much involved in natural processes as he is in supernatural events. Creation itself provides unmistakable evidence of God's handiwork.

This fourth thesis moves from the doctrine of Scripture to the doctrine of God. Any conversation about origins involves assumptions about who God is, what the world he has made is like, and how God interacts with this world. Often these assumptions lie just beneath the surface of our thinking and conversation, unarticulated or under-articulated, perhaps not fully understood or readily exposed. And yet these assumptions in turn shape conversations in sometimes unhelpful ways.

One of the main assumptions that enters into these conversations is a view of God that is more *deistic* than theistic. In our secular age, even Christians are accustomed to viewing the world in mechanistic or materialistic ways—so that while we find it quite easy to affirm that God is involved in raising someone from the dead, we just as easily slip into patterns of thinking that have little place for God in the routine workings of nature, like the rotation of the stars or the formation of clouds or the grass as it grows. That's just nature doing its thing.

This implicit naturalism hinders engagement in these discussions and limits our theological imagination in unhelpful ways. According to Scripture, God is just as much involved in the rising of the sun as he is in the parting of the Red Sea. Sure, the one may be an unusual and powerful sign of God's saving purposes, but Christians should avoid thinking that some things are "natural" processes that God isn't involved with.

To be more specific, we need to avoid being essentially atheistic in the way we view the "natural" world, as though God isn't involved in all the processes scientists like to study—things like cell divisions, photosynthesis, or condensation. As Karl Barth says of God's providential interaction with his creation, "He co-exists with it actively, in an action which never ceases and does not leave any loopholes."[7] Or consider Psalm 104, which celebrates God at work in virtually everything.

An upshot of this is that creation itself provides unmistakable evidence of God's handiwork. As the psalmist declares, "The heavens are telling the glory of God" (Ps 19:1 NRSV). Or as the apostle Paul puts it, God's "invisible attributes, namely, his eternal power and divine nature, have been clearly perceived, ever since the creation of the world, in the things that have been made" (Rom 1:20 ESV).

5. Adam and Eve were real persons in a real past, and the fall was a real event with real and devastating consequences for the entire human race.

Thesis five is likely to be a sticking point for some. An increasing number of evangelical evolutionary creationists are giving up belief in Adam and Eve as real persons in a real past.[8] The genetic evidence, at least as we now understand it, makes belief in an original human pair doubtful if not impossible; or it is at least doubtful that an original pair are the *biological progenitors* of the entire human race. Whether Adam and Eve might serve as "federal heads" of humanity in some theological sense is an option some are considering.[9]

I suspect in twenty years' time, support for Adam and Eve as real persons in a real past will be a minority view even within evangelicalism. Of course, I'm neither a prophet nor the son of a prophet, but we do appear to be on this kind of a trajectory. Should this come to pass, I for one am confident that the Christian faith will survive even though this will require some reconfiguration of our deepest convictions. Thus, we should assume a spirit of openness and a posture of optimism about how our theological understanding can and will be enriched by further engagement with science.

That being said, I personally don't find the genetic evidence compelling enough at this point to jettison belief in a real Adam and Eve in a real past.

---

[7]Karl Barth, *Church Dogmatics* III/3, ed. G. W. Bromiley and T. F. Torrance, trans. G. W. Bromiley and R. J. Ehrlich (Edinburgh: T&T Clark, 1960), 13.

[8]I'm indebted to John Walton for helping me reframe the issue in terms of Adam and Eve as "real persons in a real past" as opposed to "historical Adam."

[9]See the helpful discussion of both the scientific material and possible theological "models" in Denis Alexander, *Creation or Evolution: Do We Have to Choose?* (Oxford: Monarch, 2014), 252-304.

I admit that the evidence is mounting and at this stage looks (to my un-trained eye) impressive. But two scriptural convictions keep me tethered to the historic Christian conviction about the original human pair. The first is the testimony of Scripture, especially Adam's presence in genealogies (Gen 5; Lk 1) and in Paul's Adam-Christ typology in Romans 5. Even more com-pelling is the idea that the Christian view of salvation appears to hinge on the doctrine of original sin and the fall as an event, which in turn requires a real person to have transgressed and thus plunged humanity into a state of sin from which it needs redemption.

Much of the fabric of Christian doctrine, at least soteriology, would seem to unravel if the events described in Genesis 1–3 don't entail what Christians have historically thought they entailed. It may be the case that faithful Christians will develop biblically legitimate and theologically sensible ways of explaining the gospel apart from a real Adam and Eve. But until that point I think it is the better part of wisdom to maintain a spirit of *engaged conservatism* on this issue.

6. *Human beings are created in the image of God and are thus unique among God's creatures. They possess special dignity within creation.*

Modern science has demonstrated that there is strong biological conti-nuity between human beings and all other animals. Human beings, for ex-ample, share 98.5 percent of their DNA with chimpanzees.[10] Other lines of evidence come from the fields of comparative anatomy, embryology, and biogeography and of course the fossil record.

It is increasingly difficult, then, to claim that human beings are *qualita-tively* distinct from the animal kingdom. To quote the famous Harvard pa-leontologist George Gaylord Simpson,

> In the world of Darwin man has no special status other than his definition as a distinct species of animal: He is in the fullest sense a part of nature and not apart from it. He is akin, not figuratively but literally, to every living thing, be it an amoeba, a tapeworm, a flea, a seaweed, an oak tree, or a monkey—even though degrees of relationship are different and we may feel less empathy for our forty-second cousin the tapeworm, than for, comparatively speaking, brothers like the monkeys.[11]

---

[10]Cf. Jerry Coyne, *Why Evolution Is True* (New York: Penguin, 2009), 195.
[11]George Gaylord Simpson, *This View of Life: The World of an Evolutionist* (New York: Harcourt, 1964), 12-13.

For Christians there is much to resist about this vision. Yet it is perhaps surprising to note how much emphasis the Genesis creation account places on the continuity between human beings and other creatures. When God created human beings, he didn't cause them to fall from the sky but formed them from the dust of the earth. Adam and Eve weren't teleported to Eden from another planet. In fact, we weren't even created on our own day but share the same primeval sixth-day birthday with the rest of the creepy crawly things—toads, aardvarks, walruses, and rhinos.

Human beings eat the same food as the rest of the terrestrial creatures (cf. Gen 1:29-30). We also share the same "breath of life," are formed from the same ground, and are *nepeŝim* like the rest of animal kind (cf. Gen 1:20; 2:7). Again, it's surprising to see how much Genesis emphasizes the continuity between human beings and the rest of God's creatures.

And yet this same scriptural text clearly intends to say that something special took place on the sixth day of creation when God created human beings. The change of language is indication enough: from "Let the water teem" (Gen 1:20) and "Let the land produce" (Gen. 1:24) to "Let us make" (Gen 1:26). This is a way of indicating that God is especially involved in this final act of creation. Here the creation reaches a new stage, a high point, and God leans into the creation of humanity in a way that is distinct from what has gone before. Human beings are thus unique.

The Christian tradition has tended to locate this uniqueness in the doctrine of the *imago Dei*, or image of God. Defining what precisely this image of God entails has been vexing for theologians. But the basic point is straightforward enough—humanity is endowed by God with a special dignity. While there is continuity between humans and the rest of animalkind, the creation account implies that this sixth-day creation called "humankind" is unique.

7. There is no final conflict between the Bible rightly understood and the facts of science rightly understood. God's "two books," Scripture and nature, ultimately agree. Therefore Christians should approach the claims of contemporary science with both interest and discernment, confident that all truth is God's truth.

Every Christian, regardless of her position on the question of origins, should be able to cheerfully embrace thesis seven. While some take issue with the notion of God's "two books," the book of Scripture and the book of nature, there is good precedent for doing so. Use of the metaphor goes back at least to Augustine and can be found in esteemed places like the Belgic Confession. Using the metaphor of God's two books has helped thoughtful Christians for centuries understand the relationship between what we find in God's world and in God's Word.

The point of the metaphor is that these two books, Scripture and nature, ultimately agree. They aren't ultimately in conflict. At times in history we have thought they disagreed or were in conflict. This is because both the book of Scripture and the book of nature require interpretation. Today we want to affirm that all truth is God's truth—wherever you find it, whether in the Bible or in the creation.

A corollary of this is that Christians should approach the claims of contemporary science with both interest and discernment. Sadly, at least in popular imagination, Christians are known less for the enthusiasm and more for their skepticism toward science. But the truth is that Christians do not need to be nervous about the findings of contemporary science—as though science might unearth a defeater to the Christian faith. It won't. It can't.

We may have to live with some tension between what we believe Scripture teaches and what we understand science to be saying. But Christians, rooted in the ultimate harmony of these two books, ought to cultivate a confident patience, even as we wrestle to make sense of the seemingly contradictory results of God's two books. Remember, now we see "in a mirror dimly" (see 1 Cor 13:12 ESV). One day, all will be made clear. We may not be able to reconcile things fully in this life, but we can trust that all is reconciled and will become clear to us in the eschaton. So we wait, in hope.

8. The Christian faith is compatible with different scientific theories of origins, from young-earth creationism to evolutionary creationism, but it is incompatible with any view that rejects God as the Creator and Sustainer of all things. Christians can (and do) differ on their assessment of the merits of various scientific theories of origins.

The neo-Darwinian assertion of people like Richard Dawkins that mutations are random and that evolution is therefore necessarily unguided or blind is a metaphysical add-on to the scientific theory of evolution, not a part of the theory itself. It's a supposition derived not from any science but from a naturalistic worldview, which regrettably is thought by many to be inseparable from the science of evolutionary biology. Christians justifiably object therefore to evolutionary science being used as a *pretense* for making grand philosophical claims about the nonexistence of God or the nature of the world or what it means to be human, something we see done with an irritating amount of frequency these days. Furthermore, Christians are quite right to object to the science classroom being used as a pulpit for naturalism, not least at the taxpayer's own expense!

Yet we must understand that the supposed conflict between Christianity and evolution is more apparent than real. Alvin Plantinga, perhaps the world's leading Christian philosopher, has developed this case at length in his recent book, *Where the Conflict Really Lies: Science, Religion, and Naturalism*. He argues that conflict with evolution *as a scientific theory* is superficial, not real. Instead, the conflict really lies at the level of philosophy and theology, not science. "There is no real conflict between theistic religion and the scientific theory of evolution. What there is, instead, is conflict between theistic religion and a philosophical gloss or add-on to the scientific doctrine of evolution: the claim that evolution is *undirected*, unguided, unorchestrated by God (or anyone else)."[12]

The idea of *unguided* evolution is indeed incompatible with Christian theism. Within the biblical worldview, nothing is random. Not even a sparrow falls to the ground apart from the will of God (Mt 10:29). If in fact God created the biological diversity we see through mutation and natural selection, then he superintended the process every single step of the way. Evolution would thus be a thoroughly directed process, the means by which God has chosen to bring about life throughout history. As the conservative stalwart and biblical scholar B. B. Warfield pointed out over a century ago, evolution cannot "act as a substitute for creation, but at best can supply only a theory of the method of divine Providence."[13]

The Christian faith, in principle, is at odds not with evolution as a *science* but with evolution as a *worldview*. Christians can and do assess the merits of the science of evolution differently. That's all good and well.

---

[12]Alvin Plantinga, *Where the Conflict Really Lies: Science, Religion, and Naturalism* (Oxford: Oxford University Press, 2012), xii (emphasis original).

[13]Benjamin B. Warfield, "On the Antiquity and Unity of the Human Race," *Princeton Theological Review* 9, no. 1 (1911): 1.

But to claim that evolution is by its very nature opposed to Christianity is simply overreaching—it's not defensible philosophically or theologically. As conservative theologian A. A. Hodge argued in an essay published in 1881, less than thirty years after the appearance of Darwin's *Origin of Species*,

> Evolution considered as the plan of an infinitely wise Person and executed under the control of His everywhere present energies can never be irreligious; can never exclude design, providence, grace, or miracles. Hence we repeat that what Christians have cause to consider with apprehension is not evolution as a working hypothesis of science dealing with facts, but evolution as a philosophical speculation professing to account for the origin, causes, and ends of all things.[14]

Some Christians believe that God created the world several thousand years ago. They see this as the plain reading of Scripture and what Christians have believed for centuries. There are others who take the Bible just as seriously but see the scientific evidence a little differently and think the world is very old—several billion years. Some of these folks are sympathetic to God using evolution to bring about all the biological diversity we see (including humans), and others are not. Here's the bottom line: Christians can and do differ on their assessment of the merits of contemporary science. This is okay. We want to give one another space to wrestle with these issues. What is not okay, or what is not a Christian view, is to exclude God from the process in any way. If the earth is young, then God made it young. If the earth is old, then God made it old. If human beings came from literal dust, then God did it. And if human beings share common ancestry with other species, then God did that too.

9. Christians should be well grounded in the Bible's teaching on creation but always hold their views with humility, respecting the convictions of others and not aggressively advocating for positions on which evangelicals disagree.

Christians have both an obligation and an opportunity to be well grounded in the Bible's teaching on creation. The vision of "mere creation" advocated in this essay should not be taken by anyone as permission to

---

[14]A. A. Hodge, introduction to Joseph S. Van Dyke, *Theism and Evolution* (New York: A. C. Armstrong & Son, 1886), xx–xxii. Reprinted in Mark A. Noll, ed., *The Princeton Theology, 1812–1921* (Phillipsburg, NJ: P&R, 1983), 235.

become lax in the pursuit of deeper understanding. To achieve unity in our faith we should not have to slump to the lowest common denominator.

Christians should know what they believe about the doctrine of creation. We should also be aware of the issues of contemporary science. This doesn't require graduate work in embryology or a mastery of quantum mechanics. But Christians should aspire to know something of the state of the discussion and what other Christians are currently wrestling with at the interface of science and faith.

As we grow in the depth of our understanding of these important issues, we should mature in our ability to engage with those who hold opposing views. It is a sign of Christian maturity to be able to live with these sorts of tensions; it is a sign of childhood or adolescence to be agitated by a less than black-and-white world.

Central to this is the cultivation of the Christian virtue of humility. Sometimes we will talk about "needing humility," as though we can turn humility on like a light switch. If only it were that easy! The truth is that humility is a virtue that is only cultivated over time and with great patience and intentionality. It is also only cultivated in community, with the help and encouragement of others. This is why the apostle Paul invites Christians to work hard "to maintain the unity of the Spirit in the bond of peace" (Eph 4:3 ESV).

In practice, humility and a desire to preserve ecclesial unity mean respectfully listening to the views and opinions of others. It also means not agitating for change or grandstanding with one's own views. You may be on the side of the angels, but that doesn't give you the right—nor would it be most helpful to the body of Christ—to prove to everyone that you are in the right. On a complex, sensitive, and contentious issue like origins, it is best for evangelicals of goodwill not to aggressively advocate for positions on which evangelicals disagree.

10. Everything in creation finds its source, goal, and meaning in Jesus Christ, in whom the whole of creation will one day achieve eschatological redemption and renewal. All things will be united in him, things in heaven and things on earth.

Creation ultimately exists for Christ. He is its source, its goal, its meaning. Scripture describes Jesus with these soaring words, "He is the image of the invisible God, the firstborn of all creation. For by him all things were created, in heaven and on earth, visible and invisible, whether thrones or dominions or rulers or authorities—all things were created through him and for him. And he is before all things, and in him all things hold together" (Col 1:15-17 ESV).

Sometimes our appeals to Jesus as the "answer" can be shallow, like a trite Sunday school answer that stops conversation without actually providing answers. That's not what this final thesis is about. Instead, as Mark Noll has argued, the person of Christ provides motives for serious learning, not least in the sciences. There is a christological basis for our engagement with the doctrine of creation and the natural world. Those who confess Jesus Christ as God incarnate have the theological resources needed for a Christ-centered approach to learning.[15]

More than that, we confess that Christ is the telos of this creation. Not only its meaning but its goal—its redeemer and the source of creation's climactic resolution. Or as Scripture so pointedly says, God's will has been "set forth in Christ as a plan for the fullness of time, to unite all things in him, things in heaven and things on earth" (Eph 1:9-10 ESV).

---

[15]Mark A. Noll, *Jesus Christ and the Life of the Mind* (Grand Rapids: Eerdmans, 2011), esp. chap. 5, "'Come and See': A Christological Invitation for Science."

4

All Truth Is God's Truth

A Defense of Dogmatic Creationism

HANS MADUEME

"ALL TRUTH IS GOD'S TRUTH," Arthur Holmes argued in his classic book. Holmes reminded his generation that the living God is all-wise, eternal, and the creator of all things; his creative wisdom is "the source and norm of all truth about everything."[1] Since the wisdom of God is unchanging, his truth is also unchanging. The God of Abraham, Isaac, and Jacob knows everything that is true, whether found in Scripture or anywhere else in the cosmos. Any genuine truth that scientists discover is de facto God's truth.

For that reason, thoughtful Christians reject the myth of warfare between science and faith. The idea that Christianity and science were always enemies is simply not true to history; their historical relationship is enormously complex.[2] God is the author of special revelation *and* he is the author of nature. To the degree that any science genuinely tracks with God's creation, believers should resist the warfare myth since truth must cohere. All truth is God's truth.

Many evangelical scholars today embrace old-earth creationism, for good reason. On this standard account, the earth is roughly 4.5 billion years old (and the universe 13.8 billion years old). The scientific arguments for an

---

[1] Arthur Holmes, *All Truth Is God's Truth* (Grand Rapids: Eerdmans, 1977), 8.
[2] See John Hedley Brooke, *Science and Religion: Some Historical Perspectives* (Cambridge: Cambridge University Press, 1991).

ancient earth seem compelling, and they include evidence from radiometric dating, astronomy, cosmic microwave background radiation, the age of white dwarf stars, nucleocosmochronology, continental drift patterns, and layers of coral growth. If this consensus view is correct and all truth is God's truth, then rejecting it is no small thing. Old-earth creationism claims to hold together the Christian faith and all these lines of evidence.

A growing number of evangelicals also commend an *evolutionary* creationism.[3] Biological evolution is well supported by multiple lines of evidence and is the canonical view in modern biology. Artificial selection, paleontology, modern genetics, and embryology all point strongly to evolution and, more specifically, to *human* evolution. If the contemporary evolutionary paradigm is correct and all truth is God's truth, then evolutionary creationism has the virtue of holding together the Christian faith and all these lines of evidence for universal common ancestry.

Young-earth creationism therefore finds itself between a rock and a hard place. It contradicts a massive body of widely accepted evidence in biology, geology, and astronomy. As if that were not enough, young-earth creationism has a dismal reputation among scholars. It has enjoyed minimal academic support, historically, and what little there is leaves something to be desired. That is no surprise, since most scholars accept the scientific arguments against it; dissenters are reluctant to ruin their reputations by broadcasting their views. Even the remarkable success of creationist fundraising and coalition building among laypeople comes with telling weaknesses.[4]

The literature by young-earth creationists has been known for bad arguments. Some creationists misunderstand, or misapply, the science they are criticizing. Others project arrogance and display uncharitable, even unchristian behavior. If you have ever tried to dialogue with zealous creationists, you likely have unhappy stories to tell. Some of their most popular leaders have credentials outside the area for which they claim expertise; a few are intellectual charlatans who make misleading statements to shore up their cause. To reason with such people can be an exercise in futility.

---

[3]E.g., see Kathryn Applegate and J. B. Stump, *How I Changed My Mind About Evolution: Evangelicals Reflect on Faith and Science* (Downers Grove, IL: IVP Academic, 2016).
[4]See Ronald Numbers, "Creationism Goes Global," in *The Creationists: From Scientific Creationism to Intelligent Design*, rev. ed. (Cambridge, MA: Harvard University Press, 2006), 399-431.

Some young-earth creationists share these concerns about their movement's credibility.[5] I may rightly be accused of cherry-picking since I have ignored any positive examples in the creationist movement, but in the popular perception the shoe fits. Even among their spokespersons, who should know better, there is a tendency to treat this debate on the same level as the gospel, as if all doctrines and disagreements are equally essential, failing to recognize that good people can disagree on some of these matters.

While the situation looks grim for young-earth creationism, a different picture emerges when relevant theological factors are taken into account—or so I shall argue. But before I do so, I need to offer three points of clarification.

First, disagreements over the interpretation of the early chapters of Genesis should not be a test of orthodoxy. At no point in church history were orthodoxy and heresy defined on the basis of the age of the earth.[6] The church has always employed a notion of dogmatic rank—that is to say, not all doctrines are created equal; some are more important than others. The age of the earth, I would argue, is a key aspect of a robust doctrine of creation. However, the Trinity, the bodily resurrection of Christ, the doctrines of salvation and sin—just to name a few—these are far weightier than debates over human antiquity.[7]

Second, in the midst of science-faith controversies, one strategy is to try to meet science halfway. There can be a *limited* apologetic value to this approach. In a culture that loves to pit faith against science, some Christians are empowered when they learn that different models of evolutionary creation can accommodate much of the biblical story. Such moves dismantle the alleged Bible-science conflict and can help embattled Christians pursuing scientific callings. As an *apologetic* strategy, adjusting theological formulations to defuse tensions with science is sometimes pastorally effective (though I am less convinced about their long-term value).[8]

---

[5] E.g., see this list of bad arguments that creationists are advised no longer to use: "Bad Arguments," Creation Ministries International, accessed March 5, 2018, https://creation.com/arguments-we-think-creationists-should-not-use.

[6] That probably reflects the fact that the age of the earth was not seriously contested in the first eighteen centuries of church history. On the history of interpreting Gen 1–2, see Andrew Brown, *The Days of Creation: A History of Christian Interpretation of Genesis 1:1–2:3* (Blandford Forum, UK: Deo, 2014).

[7] At the same time, any rendering of old-earth or evolutionary creationism should be rejected if it compels Christians to jettison or undermine essential doctrines of the faith.

[8] I'm making a concession; in my view, the *truth* underlying any apologetic strategy is ultimately far more important than its pragmatic effectiveness.

Third, I am a ruling elder and theologian in the PCA (the Presbyterian Church in America). In my denomination, leaders can hold to any of the main views of creation and be "within bounds."[9] As long as they can subscribe in good faith to the Westminster Confession, young-earth, old-earth, and certain narrowly defined developmental approaches are all sanctioned ways of reading Scripture.[10] I have good friends who are far more capable representatives of my Reformed tradition and who are also old-earth creationists, and of course B. B. Warfield—a shining star in my tradition—was not opposed to evolutionary development.[11]

The rest of this chapter unfolds in two parts. I first defend the doctrine of the cosmic fall, a key theological reason that Christians like me remain inclined to young-earth creationism. Given that the cosmic fall is in deep conflict with the standard scientific picture, I then defend a view I call "dogmatic creationism," which clarifies how a Christian can still rationally hold to her position in spite of what seems to be overwhelming evidence against it.[12]

THE COSMIC FALL

Biological evolution would not be possible without the physical death of organisms. Death, arguably, is *necessary* for evolutionary development. While recent scholarship has emphasized "cooperation" and "mutuality" in the evolutionary process, competition and death remain an essential feature of the evolutionary picture.

The challenge here is twofold.

First, there is a notable tradition of Christian interpretation that views animal death as a result of the fall of Adam and Eve. Far from being an instance of patristic or medieval eisegesis, the early chapters of Genesis

---

[9]See *Report of the Creation Study Committee*, June 2000, PCA Historical Center, www.pcahistory .org/creation/report.html.

[10]PCA ministers disagree on whether the Westminster Standards teach young-earth creationism or leave the question open—e.g., see Joseph Pipa and David Hall, eds., *Did God Create in Six Days?* (Taylors, SC: Southern Presbyterian Press, 1999); Robert Letham, "'In the Space of Six Days': The Days of Creation from Origen to the Westminster Assembly," *Westminster Theological Journal* 61 (1999): 149-74.

[11]See Bradley Gundlach, *Process and Providence: The Evolution Question at Princeton, 1845–1929* (Grand Rapids: Eerdmans, 2013).

[12]In what follows I do not claim to speak for all young-earth creationists.

suggest that animal violence and death are a post-fall aberration.[13] That's the picture we see in Isaiah 11:6-9, 65:17 and 25, passages that suggest a new heaven and new earth in which there is no longer any animal suffering or death. We can infer from this glimpse of the future that, in the prelapsarian world, there was also no animal death and predation (even so, the new heaven and new earth will *exceed* Eden in glory). Animal death and meat-eating come only in Genesis 3 after the fall, consistent with the picture in Genesis 1 and 2 where humans and animals, with an exclusively vegetarian diet, dwell together peaceably.[14]

As a result of Adam's fall, God cursed the ground: "through painful toil you will eat food from it all the days of your life" (Gen 3:17; cf. 5:29). In the New Testament, the apostle Paul alludes to this passage in Romans 8:19-22 and argues that God subjected the whole creation "to frustration," in "bondage to decay" (Rom 8:20, 21). Indeed, creation has been "groaning as in the pains of childbirth right up to the present time" (Rom 8:22). The inspired apostle confirms that the "ground" in Genesis 3:17, by synecdoche, stands in for the cosmos. The effects of the fall extend wider than humanity or even earth itself—the entire cosmos is implicated.

Before the rise of geology, most believers thought that animal predation and death resulted from Adam's fall. The doctrine of a cosmic fall is no longer viable for many Christian scholars today. Young-earth creationists still defend that tradition. Old-earth creationists accept the main lines of geology and cosmology, implying millions of years of animal predation and death before any humans were on the scene (on this point evolutionary creationists agree).

Some have suggested that our intuitions about animal death and predatory violence are different from that of Scripture (you are biased because of your pet dog!). They argue that the Bible doesn't portray animal predation as bad or less than ideal. Others reinterpret the Isaiah and Romans passages.[15] Even among premodern theologians there was some hermeneutical

---

[13]Space prevents me from fleshing out a biblical understanding of death (nor does Scripture address every possible question). Minimally, Scripture assumes prelapsarian plant death, while death of higher-order animals and humans emerges in a postlapsarian setting.

[14]Admittedly, this is deeply contested; for critique of original vegetarianism, see Iain Provan, *Discovering Genesis: Content, Interpretation, Reception* (Grand Rapids: Eerdmans, 2016), 120-24.

[15]For analysis, see Michael Murray, *Nature Red in Tooth and Claw: Theism and the Problem of Animal Suffering* (New York: Oxford University Press, 2008).

diversity.[16] Exegetical caution aside, we should also guard against the danger of hermeneutical innovation as panacea, the magical cure-all whenever there is deep conflict between our current scientific explanations and what God is saying to us in Scripture. Perhaps sometimes we should *not* reinterpret Scripture; perhaps we should learn to live with the tension.

The second aspect to the challenge is *human* death. The biblical witness raises even more pressing questions here for evolutionary creationism. Scripture is adamant: human death is antithetical to the goodness of God. Romans 6:23 asserts that the wages of sin is death. According to 1 Corinthians 15:26, death is the last enemy. In the Old Testament, death is clearly a spoiler of God's good creation. Death is punishment for law breaking (see Ex 21:12; Lev 20:2, 9-13; Deut 22:21, 24). Scores of passages could be cited along these lines. The wickedness of the Canaanites and the people of Sodom and Gomorrah *justifies* their death (cf. Deut 9:5; 18:9, 12; 2 Pet 2:6; Jude 7). Both Romans 5:12-21 and 1 Corinthians 15:21-22 state unequivocally that human death results from Adam's fall.[17] Isaiah envisions the day when death will be swallowed up forever at Christ's return (Is 25:8). The first canon of the Council of Carthage (AD 418) is also noteworthy from a historical perspective: "If any one says that Adam, the first man, was created mortal, so that, whether he sinned or not, he would have died *from natural causes*, and not as the wages of sin, let him be anathema."[18]

My concern with evolutionary creationism is that it normalizes death. Both animal and human death are natural to the process of evolutionary development. In response, some Christian evolutionists have argued that when human beings were created, God set them apart from the cycle of death. Perhaps the hominids were not truly human, and then God created Adam and Eve de novo so they did not inherit that evolutionary history; then when they sinned, death entered into the human experience for the

---

[16]E.g., see Ryan McLaughlin, *Preservation and Protest: Theological Foundations for an Eco-Eschatological Ethics* (Minneapolis: Fortress, 2014), 30-32.

[17]In light of Paul's emphasis on *bodily* resurrection, the implied antithesis is *bodily* death. It is true that, in the wider biblical narrative, *spiritual* death is more threatening than physical death (e.g., Mt 10:28)—but, as John Murray observed, "the separation of the body and spirit and the return to dust of the former had far more significance [for Paul] as the epitome of the wages of sin than we are disposed to attach to it" (*The Epistle to the Romans*, vol. 1 [Grand Rapids: Eerdmans, 1959], 181).

[18]Henry Bettenson, ed., *Documents of the Christian Church*, 2nd ed. (Oxford: Oxford University Press, 1963), 59 (my emphasis).

first time. This view may not be prevalent today, but it is logically possible and avoids the hamartiological problem of human death.[19]

But a growing number of evangelicals speculate that Adam and Eve, understood as a couple or as a tribe, *evolved* from earlier ancestors; at the point of transition, God supernaturally implanted a soul in two of them. Or, perhaps, God established a special relationship with two of them—there are many other scenarios along these lines. In my judgment, they all contradict the biblical testimony by denying that human death results from the fall. After all, since Adam and Eve on these views were the product of evolution, they inherited the very processes of death from their ancestors. If God created Adam and Eve by means of evolution, once they became *Homo sapiens* (or "*Homo divinus*") they would *already have been dying*. That violates the bond between the fall and human death.

One possible way out is to say that, as soon as Adam and Eve evolved into being, God supernaturally prevented them from inheriting mortality. That move seems ad hoc and runs counter to the text, but at least it makes sense. But if the idea is that God has to intervene supernaturally in order to prevent Adam and Eve from inheriting biological mortality, that already presupposes sin and its effects. In the biblical narrative, however, the original creation was eminently "good," reflecting the very goodness of God (Gen 1:10, 12, 18, 21, 25, 31)—as one theologian remarks, "The world was not created with the Fall in prospect, still less with the curse already let loose."[20]

This debate is much ado about nothing, according to some evangelical scholars like John Walton. We should not think that *mortality* as such is a result of the fall anyway. In this view, Adam and Eve were *created* mortal. They did not die at first because they ate regularly from the tree of life; once cast out of Eden, they succumbed to their innate mortality. Only in that sense does death result from Adam's transgression. To be sure, whether Adam and Eve were mortal or immortal before the fall is a complex question in the history of interpretation.[21] However, I worry that the "Adam-created-mortal"

---

[19]Most old-earth creationists accept animal death before the fall, but they typically argue that human death came *after* the fall.

[20]Nigel Cameron, *Evolution and the Authority of the Bible* (Exeter: Paternoster, 1983), 66.

[21]In premodern and Reformation traditions, Adam and Eve's immortality implied eternal life on earth or—after a temporary period—translation into heaven, without death (like Enoch). E.g., see John Calvin's remarks on prelapsarian Adam: "In his body there was no defect, wherefore he

position normalizes death. Instead, I would argue that before the fall Adam was *conditionally* immortal just as he was *conditionally* sinless. Had he not sinned, he would have remained immortal and sinless; because he sinned, he made himself mortal and a sinner.

If we take the doctrine of the cosmic fall together with the most natural reading of Genesis 1 and 2, that is, creation took place in six ordinary days; and if we include the biblical genealogies, which, I would argue, are most consistent with a young-earth perspective (not only in light of Genesis 1 and 2, but also in light of passages like Matthew 19:4 which assume that Adam and Eve were brought into being very near the start of God's creation of all things); and finally, if we consider that a young-earth understanding was the dominant view of the premodern church—taking *all* of that together, there is a strong case for a young-earth position. This is in no way exegetically or theologically airtight; far from it, most of my argument is deeply contested.[22] My only conclusion at this juncture is that young-earth creationism has warrant in Scripture and catholic tradition (contrary to some revisionist claims).

This preliminary conclusion is at great odds with the understanding of almost all mainstream scientists today. Mind you, there is no atheistic conspiracy here. Many of them are faithful believers. There are good scientific reasons that an ancient earth and evolution are widely accepted today. Some might therefore argue that if the exegesis is inconclusive between young-earth and other perspectives, then the scientific consensus should tip the hermeneutical balance *against* any young-earth stance. Though appealing, that move merely begs the question. The only reason that the exegesis is "inconclusive" is precisely because of the prevailing scientific understanding of the world. We argue in circles if we allow that same consensus to determine what exegetical option to take.[23]

---

was wholly free from death. His earthly life truly would have been temporal; yet he would have passed into heaven without death, and without injury" (*Commentary on the First Book of Moses, Called Genesis,* trans. John King, vol. 1 [Grand Rapids: Eerdmans, 1948], 121). For a modern contrast, see Konrad Schmid, "Loss of Immortality? Hermeneutical Aspects of Genesis 2–3 and Its Early Receptions," in *Beyond Eden: The Biblical Story of Paradise (Genesis 2–3) and Its Reception History,* ed. Konrad Schmid and Christoph Riedweg (Tübingen: Mohr Siebeck, 2008), 58–78.

[22]E.g., for insightful analysis from a very different perspective, see Mark Harris, *The Nature of Creation: Examining the Bible and Science* (New York: Routledge, 2014).

[23]As Bill Dembski remarks in *The End of Christianity: Finding a Good God in an Evil World* (Downers Grove, IL: InterVarsity Press, 2009), 206n19, "If science is not the whole reason for questioning the traditional reading, it is the reason for the other reasons."

Nothing in my argument implies that our presuppositions about modern science should never impinge on exegesis. In fact, it is *impossible* to approach Scripture as a tabula rasa, a blank slate. We always come to the text with background assumptions, and many of them will be shaped by the modern scientific understanding of the world—indeed, science can sometimes rightly prompt us to discern better interpretations of the text. I only insist, in keeping with *sola scriptura*, that we submit *all* our presuppositions to Holy Writ.

DOGMATIC CREATIONISM

How then should we proceed? Let Christology be our guide. The cornerstone of Christian faith is the life, death, and resurrection of Jesus. As the apostle Paul says, "If Christ has not been raised, our preaching is useless and so is your faith" (1 Cor 15:14). However, every school child knows that resurrection is scientifically impossible (at this juncture I'm assuming a *methodologically naturalistic* science, the view held by most scientists and one that I shall question momentarily). Every known science points unequivocally against bodily resurrection. But if it is paramount for Christians to dissolve any conflicts between science and theology, then why don't they abandon belief in the resurrection? Why indeed!

An obvious answer is that the entire biblical story unravels. Take away the resurrection and we lose our union with Christ, the forgiveness of sins, justification, regeneration, the new heaven and new earth, and virtually everything central to biblical Christianity. No, thank you.

But there is a more basic reason Christians should never relinquish belief in resurrection. Our confidence in the resurrection is not grounded in the presence or absence of scientific "proof"; our confidence rests on divine revelation. As believers appropriate God's Word, the Holy Spirit instigates a confidence that these words are divinely inspired, a theological confidence that cannot be invalidated by modern science or any other aspect of human reason. The confidence is rooted in the trustworthiness of God.

No one should doubt the value of historical investigation and archeological research, but they are not the ultimate basis on which Christians trust Christ for salvation. If a global network of very smart PhDs were to develop historical arguments against the resurrection, that would be no reason for Christians to abandon belief in this central doctrine. Such arguments should not be ignored; they are potential negative defeaters, but they will never be decisive.[24]

The question then is this: Is there any scientific evidence that could possibly defeat our belief in the resurrection? Logically, yes; the resurrection is an event that happened within our space-time historical context—in principle, it is possible that science or historical investigation could disprove it. But given what we know from divine revelation, it is actually impossible. Christians can be confident that Jesus rose from the dead. No scientific counterclaims will ever convince the believer.

Not all Christian beliefs are as central or important as the resurrection of Christ. There's the rub. What should believers do when a doctrinal commitment conflicts with a well-attested scientific claim? What options do we have? I have suggested that young-earth creationism has theological and exegetical merit, but *scientifically* that view seems monumentally implausible. These four questions crystallize what is at stake:

1. How clearly is the belief warranted in the Bible?[25]

2. How central is it to the fabric of faith?

3. Is it widely believed throughout church history?

4. What is the epistemic threshold for a potential defeater to *overturn* a dogmatic conviction?

In my estimation, there is good warrant for young-earth creationism in Scripture, though by no means is it incontestable. Surely the age of the earth is not the most central tenet of the faith, but neither is it unimportant. There is good precedent in church tradition, and yet the age of the earth was never formally enshrined in creedal or conciliar statements. More could be said, of course, and Christians disagree about each of these four items. For present purposes, let us focus on the threshold criterion. At what point must we conclude that a scientific claim *passes the threshold*, that biblical texts should

---

[24]Nothing I have said invalidates the historicity of the resurrection (see 1 Cor 15:6!) or significant apologetic work defending it. My point is that the Christian's *ultimate* warrant for believing the resurrection is the intratextual—*not* extratextual—evidence of God's Word.

[25]For my use of the term *warrant*, I'm drawing on Alvin Plantinga—e.g., see his *Warranted Christian Belief* (Oxford: Oxford University Press, 2000).

be reinterpreted; at what point must we conclude that a particular doctrinal belief such as young-earth creationism needs to be abandoned?[26]

Christians have fundamentally different intuitions here. Evolutionary creationists believe that we have reached that threshold for evolution.[27] Old-earth creationists believe that we have reached that threshold for the age of the earth but not for biological evolution. Plausible arguments are made on both fronts, but I have given reasons why I am not persuaded. My question is whether it is even *rational* to maintain a belief in a young earth in spite of the overwhelming consensus.

I think so. On my view, belief in young-earth creationism is ultimately grounded in special, not general, revelation. The epistemic reasons for holding to the position are primarily exegetical and theological. My warrant for believing that Scripture yields a young-earth creationist account lies in a prior judgment about the nature of Scripture itself, Scripture as the Word of God, the canon of the church. Reliable scientific insights *can* also add to my warrant for young-earth creationism, especially if they offer additional reasons to think that the position is true. But that's not how things stand with current scientific thinking; those scientific conclusions may still function as negative defeaters, but not as warrant. They are not the positive basis for holding to the position in the first place. Let us call this position *dogmatic creationism*.[28]

If the scientific consensus is compelling enough, it *may* rightly overturn dogmatic creationism. To which someone might respond: "Surely it is compelling enough! How much more evidence do you want?" A fair question. The driving insight of dogmatic creationism is this: just because Christians have not been able to come up with scientific models that are better and more plausible than current opinion *does not mean there are no such models*.[29]

---

[26]This last question is not unrelated to the other three: the more warrant in Scripture, the more central the belief, and the more widely attested in the tradition, then the higher the threshold.

[27]On the epistemic virtues of evolutionary theory, see Gijsbert van den Brink, Jeroen de Ridder, and René van Woudenberg, "The Epistemic Status of Evolutionary Theory," *Theology and Science* 15, no. 4 (2017): 454-72.

[28]Dogmatic creationism has methodological similarities to "skeptical theism"—e.g., see Michael Bergmann, "Skeptical Theism and the Problem of Evil," in *The Oxford Handbook of Philosophical Theology*, ed. Thomas Flint and Michael Rea (Oxford: Oxford University Press, 2009), 375-99.

[29]Most young-earth creationists would insist that, in at least some cases, they *have* developed models with greater explanatory and predictive power than conventional theories. While dogmatic creationism does not contest that claim, it adopts a worst-case scenario—even if *none* of the current creationist models succeeds, that hardly rules out the viability of the position.

I confess that there *are* empirically rooted reasons that we should be skeptical about the consensus at those points where it conflicts with Scripture, even if I have no idea what those reasons are.

Ever since the publication of Henry Morris and John Whitcomb's *The Genesis Flood*, many young-earth creationists have tried to amass scientific arguments in favor of their position. I am pessimistic about most of those efforts.[30] So far none of them has been widely received or brought about a scientific revolution. Again, I say: just because Christians have not been able to come up with alternative models that are better and more plausible than the consensus *does not mean there are no such models.* But is this just hand waving or wishful thinking? On what grounds can I assume there are empirically rooted reasons that we should be skeptical about the consensus view when it conflicts with young-earth creationism, even if I have no idea what those reasons are?

Let me sketch five points to support the epistemic stance of dogmatic creationism.[31]

1. The doctrine of God. In a famous symposium first published in 1951, Antony Flew posed the following question to the theist: "What would have to occur or to have occurred to constitute for you a disproof of the love of, or the existence of, God?"[32] Flew suspected that belief in God is *unfalsifiable*; there seems to be "no conceivable event or series of events the occurrence of which would be . . . a sufficient reason for conceding 'there wasn't a God after all' or 'God does not really love us then.'"[33] In his response to Flew, Basil Mitchell, former professor at Oxford, admitted, "It is true that [the theologian] will not allow it—or anything—to count decisively against it, for he is committed by his faith to trust in God. His attitude

---

[30]Noteworthy exceptions include work by Todd Wood, Kurt Wise, Paul Garner, and others; if they can strengthen the plausibility of young-earth models of natural history, then their project supports the more theological focus of dogmatic creationism. E.g., see Todd Wood and Megan Murray, *Understanding the Pattern of Life: Origins and Organization of the Species* (Nashville: B&H, 2003); Kurt Wise, *Faith, Form, and Time: What the Bible Teaches and Science Confirms About Creation and the Age of the Universe* (Nashville: B&H, 2002); Paul Garner, *The New Creationism: Building Scientific Theories on a Biblical Foundation* (Darlington, England: EP Books, 2009).

[31]My five points are not exhaustive.

[32]Antony Flew, "Theology and Falsification," §A, in *New Essays in Philosophical Theology*, ed. Antony Flew and Alasdair MacIntyre (London: SCM, 1955), 99.

[33]Flew, "Theology and Falsification," 98.

is not that of the detached observer, *but of the believer.*[34] Mitchell then offered a parable,[35] which deserves a lengthy citation:

> In time of war in an occupied country, a member of the resistance meets one
> night a stranger who deeply impresses him. They spend that night together in
> conversation. The Stranger tells the partisan that he himself is on the side of the
> resistance—indeed that he is in command of it, and urges the partisan to have
> faith in him *no matter what happens.* The partisan is utterly convinced at the
> meeting of the Stranger's sincerity and constancy and undertakes to trust him.
>
> They never meet in conditions of intimacy again. But sometimes the
> Stranger is seen helping members of the resistance, and the partisan is grateful
> and says to his friends, "He is on our side."
>
> Sometimes he is seen in the uniform of the police handing over patriots
> to the occupying power. On these occasions his friends murmur against him:
> but the partisan still says, "He is on our side." He still believes that, in spite of
> appearances, the Stranger did not deceive him. Sometimes he asks the
> Stranger for help and receives it. He is then thankful. Sometimes he asks and
> does not receive it. Then he says, "The Stranger knows best." Sometimes his
> friends, in exasperation, say, "Well, what *would* he have to do for you to admit
> that you were wrong and that he is not on our side?" But the partisan refuses
> to answer. He will not consent to put the Stranger to the test.[36]

Mitchell goes on to say that the soldier (i.e., "partisan") takes the claim
"The Stranger is on our side" as a basic truth *that cannot ultimately be contra-
dicted.* The soldier takes this position because of his unwavering trust in the
Stranger. But he of course "recognizes that the Stranger's ambiguous be-
haviour *does* count against what he believes about him. It is precisely this
situation which constitutes the trial of his faith."[37] Now, what if the soldier asks
for help but does not receive it? He can conclude one of two things: (1) the
Stranger is not on our side, or (2) the Stranger is on our side but has his
reasons for withholding help. The soldier keeps refusing the first choice. He
will never conclude the Stranger is not on his side. Mitchell then asks, "How
long can he uphold the second position without its becoming just silly?"[38]

---

[34]Flew, "Theology and Falsification," 103, my emphasis.
[35]Antony Flew's original article, to which Mitchell is responding, also included a parable.
[36]Basil Mitchell, "Theology and Falsification," §C, in Flew and MacIntyre, *New Essays*, 103-4.
[37]Mitchell, "Theology and Falsification," 104.
[38]Mitchell, "Theology and Falsification," 104.

Mitchell responds that the soldier cannot answer that question in advance; it depends on factors like the strength of the initial impression made by the Stranger, the way in which the soldier assesses the Stranger's ambiguous behavior, and so on. But Mitchell insists that the soldier's belief about the trustworthy character of the Stranger *qualifies as a sufficient explanation*. The soldier's position has an "irreducible circularity" that makes sense in light of *the character* of the Stranger.[39]

This parable illuminates dogmatic creationism. I am not making the implausible claim that the age of the earth is as central to the faith as the existence of God. Instead, I wish to draw attention to the nature of God and our trust in him. If the believer is convinced that young-earth creationism is revealed in Scripture, then that dogmatic commitment necessarily involves the very nature of God: Does he lie? At bottom, do we trust him? The scientific consensus against young-earth creationism is powerful, but not irrefutably so. My belief that there are better empirically rooted models is warranted, even if I currently have no idea what they might be.

2. The fallibility of science. I am a scientific realist. I think science aims at true descriptions of the world. But one lesson that anti-realists have taught us is that science is a fallible enterprise. Many of our best theories are underdetermined by the data used to support them. Furthermore, the history of science reveals many widely held scientific beliefs that we now know were false (even though they were often remarkably successful in their day)—for example, geocentrism, phrenology, the ether, and the four humors. We have historians like Thomas Kuhn to thank for blowing the whistle.

Given this history of science, there may well be alternative empirically rooted models about our origins that are better explanations of the data that we have. Dogmatic creationism takes the view that there is at least one such model. That we do not currently know what it is has to do with our human finitude and possibly the effects of sin on our cognitive processes.[40]

---

[39]Nigel Cameron, *Biblical Higher Criticism and the Defense of Infallibilism in 19th Century Britain* (Lewiston, NY: Edwin Mellen, 1987), 379n18.

[40]The mere recognition of scientific fallibility, on its own, is likely insufficient—after all, my sense experience is fallible, but it would take a great deal to negate my belief that I'm awake as I write this sentence! The salient question, for dogmatic creationism, is whether I'm *justified* in believing that relevant scientific counterclaims are unreliable. My thanks to James Anderson for pressing me on this point.

3. The noetic effects of sin. Sin has affected our intellectual capacities; hence we speak of the "noetic" effects of sin (from Greek *nous*, "mind"). The scientific enterprise is enmeshed in the noetic effects of sin. These vary widely across the scientific disciplines; for instance, sin's noetic effects on chemistry and physics may be less than its effects on psychology.[41] As Graham Cole writes, "Every discipline presupposes some doctrine of the human. In some disciplines that doctrine is very much on the surface and potential conflict between the Christian and others will be more to the fore. One might suggest that there is a principle of proximity to the anthropological."[42] This reality is relevant to the merits of dogmatic creationism.

To be fair, the noetic effects of sin also affect biblical interpretation (consider the countless Protestant denominations!). However, I have one caveat. The Bible comes to us infallibly in the form of human language. There is a sense in which the data of nature are "inerrant," but science can only access those data through fallible linguistic formulations (our theorizing).[43] So while Scripture and nature are both *revelatory*, they differ in one key respect. God promises that his Holy Spirit will guide believers as they interpret Scripture, but he has made no such promises with the interpretation of nature. That means there is a fundamental asymmetry between biblical interpretation and scientific theorizing. Constantly stressing the interpretative dimension of Scripture, as if it were on a par with scientific theorizing, threatens the clarity of Scripture and invites exegetical skepticism. As scientific theories encroach on areas formerly thought to be within the biblical purview, we become increasingly skeptical that the relevant passages have *any* revelatory status—or if they do, that we can know them.

I do not endorse a radical fideism that ignores everything science tells us. I take science seriously, but I take Scripture more seriously. On the one hand,

---

[41]See Stephen Moroney, *The Noetic Effects of Sin: A Historical and Contemporary Exploration of How Sin Affects Our Thinking* (Lanham, MD: Lexington, 2000); Alvin Plantinga, "Sin and Its Cognitive Consequences," chap. 7 in *Warranted Christian Belief*.

[42]Graham Cole, "Scripture and the Disciplines: The Question of Expectations," *Zadok Papers* 142 (2005): 5.

[43]I'm assuming for the sake of argument that the "data" of nature are best construed *as* God's general revelation. But that assumption is questionable; properly speaking, general revelation involves the attributes *of God*, not the natural data of creation. E.g., see Nicolaas Gootjes, "General Revelation and Science: Reflections on a Remark in Report 28," *Calvin Theological Journal* 30 (1995): 94-107.

Scripture must be interpreted; Scripture and theology are not Siamese twins. On the other hand, we should resist relativizing Scripture by placing it on a level epistemic playing field with scientific interpretations of nature. That move collapses the very idea of biblical authority, since the consensus among scientists at any given historical moment delimits what is exegetically possible. As the explanatory reach of accepted scientific theories expands, the epistemic jurisdiction of Scripture shrinks into an increasingly small area of "spiritual" truth.

4. The devil and the powers of darkness. The devil and his minions are exiled from the Western conception of reality. In Rudolf Bultmann's famous line, "It is impossible to use electric light and the wireless and to avail ourselves of modern medical and surgical discoveries, and at the same time to believe in the New Testament world of spirits and miracles."[44] But such views should not be taken seriously. The dark reality of demonic powers was never doubted in church history—indeed, the Son of God came to destroy the works of the devil (1 Jn 3:8).[45] Some conflicts between science and theology are not *merely* conflicts between ancient doctrines and new scientific perspectives. Other forces are bent on eviscerating the power of divine revelation. I have no private knowledge into demonic entities and their abilities, but Scripture attests that there is far more to this world than meets the natural eye.

The New Testament witness to demon possession and exorcisms is pervasive (e.g., Mt 8:28-34; 15:21-28; 17:14-21; and many such passages). But there are other clues to a broader supernatural drama. Satan is described as "the prince of this world" (Jn 12:31) and "the god of this age" (2 Cor 4:4); according to Ephesians 6:12, "our struggle is not against flesh and blood, but against the rulers, against the authorities, against the powers of this dark world and against the spiritual forces of evil in the heavenly realms" (see also Eph 2:2). Such texts reveal that the devil and other fallen angels, though defeated by Christ on the cross, have considerable sway over human society (e.g., see Jn 17:15).

The individuals doing research at scientific institutions are not removed from these demonic realities, even if we are unable to specify exactly where diabolical influence extends. While Scripture is relatively silent on details, we can be certain there is much we do not know. In light of what God has revealed,

---

[44]Rudolf Bultmann, *Kerygma and Myth*, ed. Hans Bartsch, trans. Reginald Fuller (New York: Harper & Row, 1961), 5.

[45]Cf. Jeffrey Russell, *Satan: The Early Christian Tradition* (Ithaca, NY: Cornell University Press, 1981).

the powers of darkness play a role in conflicts of all kinds, especially those that harm Christian faith and witness. This extends to the debates Christians have about origins. That fact increases the warrant for dogmatic creationism, even if we have no way of knowing the extent of that influence.[46]

5. *The problem of methodological naturalism.* The scientific theories lined up against young-earth creationism are often underwritten by methodological naturalism. Such theories cannot appeal to direct or supernatural divine action (including divine revelation). Scientific theories can operate only within the natural, causal nexus of the world. If we take the evidence base to be "the set of beliefs to which I appeal to in conducting an inquiry,"[47] then methodological naturalism consistently limits the evidence base from which scientists can draw. Since the ultimate warrant for holding to creationism lies in Holy Scripture, scientific counterarguments can serve as potential defeaters. But given the self-imposed myopia of methodological naturalism, other legitimate factors have been eliminated from the evidence base, rendering the scientific conclusions less persuasive.[48]

Having sketched out these five points, let me circle back to my original question: What is the epistemic threshold for a potential defeater to *overturn* a dogmatic conviction? If we mean "central" doctrines of the faith *that are clearly attested in Scripture*, science *cannot* overturn such doctrinal interpretations. Scripture cannot be broken (cf. Jn 10:35). In the case of more "peripheral" doctrines that are not as clearly attested, scientific theories can be potential defeaters, *but the threshold would be very high.*

When do we know we've reached the threshold? This question is difficult to answer in the abstract. In my view, heliocentrism passes the threshold; evolution does not.[49] But how do I justify that conclusion? My short answer is that scientific theories that are closer to the observational data are more

---

[46]Young-earth creationists are not immune to the powers of darkness, but that is not my focus here.

[47]Alvin Plantinga, "Games Scientists Play," in *The Believing Primate: Scientific, Philosophical, and Theological Reflections on the Origin of Religion*, ed. Jeffrey Schloss and Michael Murray (Oxford: Oxford University Press, 2009), 160.

[48]Ronald Numbers argues that Christians, ironically, were among the first to endorse methodological naturalism—see his "Science Without God: Natural Laws and Christian Beliefs," in *When Science and Christianity Meet*, ed. David Lindberg and Ronald Numbers (Chicago: University of Chicago Press, 2003), 265-85.

[49]Proper reading of Scripture discerns that its ordinary-language (i.e., phenomenal) style does not demand geocentrism, though this hermeneutical perspective is itself the *beneficiary* of Copernican science.

reliable than those that depend on supplemental scientific theories. As hypotheses move farther away from the observational data, the likelihood of error rises. Granted, that position is still controversial and needs more nuance, but I think evolution and old-earth creationism rely on auxiliary scientific theories; by the lights of dogmatic creationism, neither position passes the threshold.

CONCLUSION

In this chapter, I gave one key argument in favor of a young earth. I also tried to defend a position I call "dogmatic creationism." On this view, young-earth creationism is ultimately grounded in Scripture. The current scientific consensus is impressive, but there are dogmatic reasons to think that better, empirically rooted models are available—even if we do not currently know what they might be.

Many Christians disagree that young-earth creationism is grounded in Scripture. I have tried to argue, however, that *if* young-earth creationism is true, then the discrepancy between *that* fact and received scientific wisdom invites several reflections, among them the doctrine of God, the fallibility of science, the noetic effects of sin, the role of demonic powers, and the problem of methodological naturalism. And let me clarify, again, that this chapter is not trying to identify which positions on creation are within the bounds of orthodoxy. Orthodoxy is broader than dogmatic creationism or any single denomination.

One closing question: What about the beautiful coherence of truth, that all truth is God's truth? I happily affirm it, recognizing that it is best understood *eschatologically*. For now, we see through a glass, darkly. At the eschaton, we will see face to face; we will see how the truths in all the disciplines cohere together and cohere in Christ. In the meantime, dogmatic creationists live with the tension. Sometimes we lack the clarity to resolve certain conflicts between science and theology. On this side of Christ's return, we may never receive that clarity, and that's okay. We walk by faith and not by sight.[50]

---

[50]I am grateful to James Anderson, Robert Erle Barham, Bill Davis, Geoff Fulkerson, Paul Garner, Paul Gesting, Tim Morris, Keith Plummer, Michael Radmacher, Ted Van Raalte, John Wingard, and Todd Wood for helpful comments on an earlier draft.

Part Two

The DOCTRINE of
CREATION EXPLORED

5

Is the World Sacramental?

Ontology, Language, and Scripture

JEREMY MANN

L E CORBUSIER, THE SWISS-FRENCH WRITER, painter, and co-designer of the United Nations Headquarters, once described a home as "a machine to live in."[1] In late modernity, this is an apt description of the whole world. Instead of serving as a reflection of a higher order, the cosmos in the twenty-first century is largely a human artifice, and the notion of a higher telos is absurd. John Milbank and coauthors write, "For several centuries now, secularism has been defining and constructing the world. . . . In its cyberspaces and theme-parks it promotes a materialism which is soulless, aggressive, nonchalant and nihilistic."[2]

Faced with this reductionistic picture, Christians have gradually mounted a resistance. Interestingly, theologians did not lead the charge. Enlivening the doctrine of creation was largely a more popular movement, championed by Christian artists, poets, pastors, and conservationists.[3] Over time, Bible scholars and theologians have put new energy and language toward reviving

---

[1] Le Corbusier and Jean-Louis Cohen, *Toward an Architecture* (Los Angeles: Getty Research Institute, 2007), 151.

[2] John Milbank, Graham Ward, and Catherine Pickstock, "Suspending the Material: The Turn of Radical Orthodoxy," in *Radical Orthodoxy: A New Theology*, ed. John Milbank, Graham Ward, and Catherine Pickstock (London: Routledge, 1999), 1.

[3] David Jones, "Art and Sacrament," in *Epoch and Artist: Selected Writing by David Jones*, ed. Harmon Grisewood (London: Faber and Faber, 1959); Robert Farrar Capon, *The Supper of the Lamb* (New York: Doubleday, 1969); Annie Dillard, *Pilgrim at Tinker Creek* (New York: Harper's Magazine Press, 1974); Wendell Berry, *The Gift of Good Land* (San Francisco: North Point, 1981).

both the relevance and the mystery of the doctrine of creation, with the Christian academy now becoming a place of focused reflection on this topic.[4] "The world is charged with the grandeur of God," begins Gerard Manley Hopkins's famous poem and a large number of Christian college course syllabi.

One element of this larger effort uses variations of the word *sacrament* to describe creation's significance. "A Sacramental approach," according to Norman Wirzba, "radicalizes our relationship to this earth by encouraging us to see the physical places in which we move and from which we live as expressions of the divine life."[5] Instead of cordoning off special sites of divine favor, a "sacramental ontology" depicts creation qua creation as a place of communion with God. In a sacramental ontology the material composition of this world—all of it, from bare rocks to baroque requiems—is a spiritual matter.[6] A definition can be taken from Hans Boersma: "[Sacramental ontology] is the conviction that historical realities of the created order served as divinely ordered, sacramental means leading to eternal divine mysteries."[7]

The primary task of this chapter is to provide a preliminary orientation for the larger discussion of a sacramental world.[8] I will disentangle the varied senses

---

[4]This interest can be seen in a number of ways, the breadth and depth of which can only be gestured at here. It includes creation as an anchoring locus for wider systematic theologies (Jürgen Moltmann, *God in Creation: A New Theology of Creation and the Spirit of God* [Minneapolis: Fortress, 1993]), creation as an integrating topic (Colin Gunton, *The One, the Three, and the Many: God, Creation, and the Culture of Modernity* [Cambridge: Cambridge University Press, 1993]), new attention to care for the natural world (Steven Bouma-Prediger, *For the Beauty of the Earth: A Christian Vision for Creation Care* [Grand Rapids: Baker Academic, 2001]), theological projects that appreciatively gather resources from other traditions' approaches to the doctrine of creation (John Webster, "Purity and Plenitude: Evangelical Reflections on Congar's *Tradition and Traditions*," *International Journal of Systematic Theology* 7 [2005]: 407), rearticulations of the divine character congruent with the created order (Randall Zachman, "'God Manifested in God's Works': The Knowledge of God in the Reformed Tradition," in *The Death of Metaphysics; The Death of Culture: Epistemology, Metaphysics, and Morality*, ed. Mark J. Cherry [Dordrecht, Netherlands: Springer, 2006], 71-97), and appetite for reflection on the manner in which art and culture natively depict eternal truths (Jeremy Bebgie, *Resounding Truth: Christian Wisdom in the World of Music* [Grand Rapids: Baker Academic, 2007]). More examples can be found in Martha L. Moore-Keish and George Hunsinger, "Twentieth-Century and Contemporary Protestant Sacramental Theology," in *The Oxford Handbook of Sacramental Theology*, ed. Hans Boersma and Matthew Levering (New York: Oxford University Press, 2015), 297-416.
[5]Norman Wirzba, *The Paradise of God: Renewing Religion in an Ecological Age* (New York: Oxford University Press, 2007), 222.
[6]The term *sacramental ontology* was first coined by Yves Congar.
[7]Hans Boersma, *Nouvelle Théologie and Sacramental Ontology: A Return to Mystery* (New York: Oxford University Press, 2009), 289.
[8]A more in-depth treatment is the task of my dissertation, still in progress.

of the term *sacramental* and explore the way the word can function to help re-enchant the world. I will then briefly scrutinize the weightiest claims in which it is put to use. The final section will offer reflections on how ministers of God's people can approach this conversation with both discernment and grace.

SHARPENING THE SACRAMENTAL PICTURE

Daniel Treier has captured well the need for a more rigorous account of sacramental ontology: "One can enthusiastically endorse recovering a sense of divine mystery regarding the Eucharist, tradition, Scripture, and indeed all of God's creatures. At that point, though, more ontological detail about such participation—or at least clarity about what that would not mean—is required."[9] As Treier acknowledges, reconsideration of the grounding of creation's goodness is something worth celebrating. Insofar as the sacramentalists prompt consideration of the nature and significance of what God has made, all Christians can benefit. Neglecting these questions comes with great peril, both in the church's theological depiction of the world and in the individual efforts of Christians to engage it.[10] The pitfalls of proto-Gnosticism and radical dualism have attended Christian practice for its whole history, and the explosion of scientific advancement in recent centuries has introduced new challenges.[11] On one telling, sacramental ontology was the rudder that guided the ark of the Christian church through stormy seas all the way from the Fathers to late-medieval Scholasticism. Beyond that era it proved itself useful through many contemporary streams of Christian thought.[12] According to this view, wider efforts to retrieve this framework should be celebrated, given the anemic condition of many congregations' understanding of God's good world.[13]

---

[9]Daniel J. Treier, "Heavenly Participation: The Weaving of a Sacramental Tapestry—A Review Essay," *Christian Scholar's Review* 41 (2011): 69.

[10]Jonathan R. Wilson offers some reflection on the problems of this "atrophied doctrine" in *God's Good World: Reclaiming the Doctrine of Creation* (Grand Rapids: Baker Academic, 2013), viii.

[11]See Phillip J. Lee, *Against Protestant Gnostics* (New York: Oxford University Press, 1993). While the title is somewhat deceiving and Lee is occasionally provocative to a fault, his work thoroughly traces both historic and recent Gnostic tendencies in North America.

[12]David Brown, "A Sacramental World: Why It Matters," in *The Oxford Handbook of Sacramental Theology*, ed. Hans Boersma and Matthew Levering (New York: Oxford University Press, 2015), 605.

[13]As Max Weber saw it, disenchantment was largely the work of northern European Protestants, a sociological reading that Charles Taylor supports (Charles Taylor, *A Secular Age* [Cambridge, MA: Belknap Press of Harvard University Press, 2007]).

Others argue that sacramental ontology was not simply assumed in Christian history, nor is it without controversy: terminology that developed to describe God's unique presence and manifest grace should not find application to the whole cosmos without some caution.[14] Furthermore, and more immediately relevant to our present concern, the mysterious character of sacramentality has made it hard to discern exactly what is being affirmed and what it entails (hence Treier's request for a more rigorous account). What also remains unclear is the extent to which sacramental ontology depends on a particular underlying theological grammar or, relatedly, the viability of a similar story told in a different idiom. Sacramental theology has not as yet "become Protestant," both in the sense that its adoption is relatively rare within the tradition and, perhaps more significantly, in the sense that its claims have not been woven into the larger fabric of Protestant theology.[15]

A large part of the challenge of clarifying sacramental language consists in the deeply embedded character of its use, combined with a uniquely wide purview. Evaluation of sacramental language requires consideration of the very grounds of existence alongside the quickly passing trends of language and varied theological emphases, themselves responsive to constantly changing cultural movements. Part of the debate itself concerns the proper location of mystery and the use of figure and analogy. David Brown maintains that even the seemingly innocuous desire for more precise definitions resulted in "obfuscating a larger vision." According to Brown, a sacramental picture is better thought of as a kind of framing background that suggests more than it specifies.[16] While this position has some warrant, the discussion in the present era is hardly one of scholastic exactitude. Some communities are suffering from a kind of "sacramania," where it appears the sacramental label is affixed with greater and greater enthusiasm the further it is found from a patient, thorough treatment of the topic. Part of the problem of understanding "sacramental" is the fact that adding an -*al* suffix on the end of a word does not yield predictable outcomes. *Economic* means "relating to economics," but *economical* generally means "inexpensive." A sacramental

---

[14]One collection of these voices is *Radical Orthodoxy and the Reformed Tradition: Creation, Covenant, and Participation*, ed. James K. A. Smith and James Olthuis (Grand Rapids: Baker, 2005).

[15]Both Milbank and Boersma, the two most emblematic Protestant figures advocating for a sacramental picture, are comfortable acknowledging certain misgivings about other aspects of Protestant thought.

[16]Brown, "Sacramental World," 605.

world could be a world that simply has sacraments in it, or it could be a world that is itself a sacrament, or it could be a world that has features in common with sacraments.

Additionally, often the people using "sacramental" are not engaged in careful dogmatic theology and are not trained as theologians. Consider the patient, learned concern of someone who makes hammers and wrenches in contrast to someone who is putting together a birdhouse and looking for a tool to help. I recently picked up a broken length of galvanized pipe, and my four-year-old son said, "That's not a hammer, Daddy." No doubt he will be ruminating on my philosophical reply for some time. When a poet says a sun-dappled forest glade is a kind of sanctuary, a sacramental space, she is trying to evoke something, often intentionally operating outside the normal bounds of theological discourse.

To consider this from a different angle, we can look at the analogies between calling the created world "sacramental" and calling certain types of embodied or contextual ministry approaches "incarnational." Is this an inaccurate or, worse, dangerous pair of descriptions? Surely both key into something helpful. Just as Christ voluntarily took the form of a servant, surrendering in some sense his status, privileges, and comfort for the sake of lost sheep, so too a missionary travels overseas or a pastor moves to a disadvantaged neighborhood. Surely only willed obstinacy prevents one from recognizing the clear similarities and the clear differences—of course a missionary does not take on a new nature or bridge whole new realms of being.

One further complicating factor bears brief mention. The historical emphases of Orthodox, Catholic, and Protestant dogma were originally a fairly reliable set of differing approaches to this topic. But in recent years these theological communities have morphed, combined, and divided in various ways, in part because of fresh recognition of the variety of sacramental theologies in the early church.[17] This makes the conceptual landscape even more complex. Further, what were once integrated systems of theology have Balkanized into various discrete units of thought, positions, and priorities

---

[17]Both books by Paul Bradshaw have been influential on this front; see Bradshaw, *The Search for the Origins of Christian Worship: Sources and Methods for the Study of Early Liturgy* (Oxford: Oxford University Press, 1992), and Bradshaw, *Reconstructing Early Christian Worship* (London: SPCK, 2009).

that can be fused together in innumerable combinations, sometimes despite hidden (or not so hidden) incongruities.[18]

For all these reasons, a systematic analysis of the sacramental is overdue. But the goal of more rigorous scrutiny is not meticulousness for its own sake. Along with its exultant praise of God, the church honors him by choosing carefully the words and concepts used to make sense of his creation, particularly in light of how often Scripture invites consideration of creation's marvels to fund worship of its Maker. Let us now consider the various ends to which sacramental language is employed.

SIX USES OF SACRAMENTAL LANGUAGE

What are we trying to do when we speak of re-enchanting the world? What role does the word *sacramental* play? Before more closely examining uses of sacramental language, let us observe their common end: establishing some manner of closeness between creation and God. Against any picture of the created world existing independent of God, unrelated to God, or disregarded by God, the sacramentalist seeks to establish and maintain a strong relationship. "Heaven is my throne, and the earth is my footstool," begins Isaiah 66, acknowledging two domains but emphasizing God's immanence in all places.[19] There are other spatial possibilities for describing this nearness. Paul Tillich, concerned that such a high-exalted God sounds too distant to modern ears, spoke of God instead as "the ground of our being," that is, that which is beneath humanity and allows it to exist and stand.[20] John Henry Newman once preached of how "every breath of air and ray of light and heat, every beautiful prospect is, as it were, the waving of the robes of those whose faces see God in heaven."[21] The realm of the divine is not high in the clouds or beneath the deepest foundations but "just behind a thin veil alongside us."[22]

---

[18]Michael Fahey, "Sacraments," in *The Oxford Handbook of Systematic Theology*, ed. John Webster, Kathryn Tanner, and Iain Torrance (Oxford: Oxford University Press, 2007), 267-84.

[19]This phrase occurs both in Is 66 and in Stephen's sermon in Acts 7.

[20]Paul Tillich, *Systematic Theology*, vol. 1 (Chicago: University of Chicago Press, 1973), 61.

[21]John Henry Newman, *Parochial and Plain Sermons*, reprint ed. (London: Longmans & Green, 1891), 453.

[22]Brown, "Sacramental World," 603-15. G. K. Chesterton takes this notion of parallel realms even further, describing his favorite color, red, as "the place where the walls of this world of ours wear thinnest and something beyond burns through." G. K. Chesterton, *Alarms and Discursions* (London: Book Jungle, 2007), 81.

Proceeding from the general heading of close interrelation, sacramental language can be more carefully distinguished by classifying any particular use—whether "sacrament" or "sacramental"—into one of six categories or senses of the term.[23] To attempt this classification exercise is not to claim that the various senses outlined do not overlap conceptually—still less that in any given use of the term an author has only one sense in mind. One might wonder if the coherence of the idea of sacramentality itself concerns its organic unity and dissecting it destroys its raison d'être. This is a legitimate objection, but the attempt to map the varied uses of sacramental language is still useful, given what A. N. Williams has deemed an "unhelpfully broad" conception of what passes under the heading.[24] A preliminary taxonomy will help clarify the diverging concerns of thinkers who look indistinguishable at first blush. It will also reveal areas of potential agreement among theologians who hold opposing views on when to use the specific term *sacramental* but who still share common ground on the underlying claims for which the word is put to use. Most broadly, a catalog of senses will aid patient evaluation of the wider phenomena. Let us now turn to the six.

First, either all creation or something particular in it can be described as a sacrament in the formal sense, that is, as a *rite* established by Christ for the spiritual edification of his people.[25] This first category is the most distinctive and warrants separate consideration. It is often linguistically differentiated from the next five senses through use of the noun form, that is, "a sacrament" or "the sacraments," in contrast to the adjectival form, "sacramental." While there are examples of the term *sacrament* functioning in uses two through six, even the strongest sacramentalist positions preserve some distinction between (1) the Eucharist, baptism, and potentially other ecclesial sacraments and (2) the varied ways by which creation exists analogously or functions

---

[23]Occasionally an author will use the term in a way that does not transparently present the intended sense. Other times a term is intentionally broad, capturing multiple senses in one expression. But I believe that all uses belong to at least one of these categories.

[24]A. N. Williams, "Nouvelle Théologie and Sacramental Ontology: A Return to Mystery," *Modern Theology* 26, no. 3 (2010): 488.

[25]The classic definition of a sacrament, "the visible sign of an invisible grace," is a formulation derived from Augustine. See E. J. Cutrone, "Sacraments," in *Augustine Through the Ages*, ed. Allan D. Fitzgerald (Grand Rapids: Eerdmans), 741-47. Augustine's original phrasing includes more dimensions: "from visible to invisible, from corporeal to spiritual, from temporal to eternal things" (*Epistles* 55.13).

sacramentally. This first sense of the term as rite is also not generally the primary concern of wider efforts to re-enchant all creation. This is because the sacraments in this first sense are much more carefully regulated by dogma and liturgical practice. As a result, the sacraments themselves have remained largely insulated from the secularizing effects of modernism, both in the nature of the threat it poses and in the way the sacraments are defended.

The second sense in which all creation or a certain part of it is said to be sacramental is in its status as an unmerited *gift*, an intentional expression of divine grace. John Milbank expresses this sense with characteristic artistry: "For in theology there are no 'givens,' only 'gifts.' . . . In creation there are only givens in so far as they are also gifts: if one sees only objects, then one misapprehends and fails to recognize true natures."[26] Creation as an act of grace confronts two positions. First, it denies the notion of pure nature conceived of by Scholastic Thomism as autonomous and devoid of grace.[27] Second, the willed giving of existence to all creation rejects the strict Neoplatonic view that creation is an emanation from God.

The third sense in which creation is described as sacramental involves its *sacred* or holy character. There is an etymological basis for this category, the Latin root *sacer*, meaning "to sanctify, make an oath, or form a treaty." *Sacrament* is derived from the Latin word *sacramentum*, which was used to translate the Greek word *mystērion* (μυστήριον) in the New Testament, most often found in Paul's letters as a previously unknown purpose of God revealed in Christ (Rom 16:25-26; Eph 3:3-4, 9; Col 1:26-27).[28]

Fourth, creation can be described as sacramental in the sense that it involves *participatory* union with God's being. John Milbank defines participation as "the notion of the cosmos as *sharing in*, displaying to a certain

---

[26]John Milbank, *Being Reconciled: Ontology and Pardon* (New York: Routledge, 2003), xi

[27]Karl Rahner's "transcendental Thomism" can be seen as a more recent Catholic argument for strong bifurcation between the natural and the supernatural.

[28]*Mystērion* had a much wider range of application than *sacramentum*, being applied to many kinds of unexplained phenomena: secret medicines, magical formulae, and obtuse philosophy teachings (Plato, *Thaetetus* 156a; *Gorgias* 497c). G. E. H. Lampe, *A Patristic Greek Lexicon* (Oxford: Clarendon), 891-93. *Sacramentum* was generally used only for oaths (particularly in the military), obligations, or initiation. Only by the fourth century was *mystērion* used commonly to refer to the rites of the church, although Echle has shown that Clement of Alexandria and Origen used the term in describing baptism earlier. H. A. Echle, "Sacrament Initiation as a Christian Mystery Initiation According to Clement of Alexandria," in *Vom Chrisliche Mysterium*, ed. A. Mayer et al. (Düsseldorf: Patmos, 1951), 54-65.

degree, the divine essence."[29] The nature of such participation is a matter of considerable debate, but the seed of the idea begins with recognition that God is the source of all that exists, and that his own being continually sustains the being of all creation. Furthermore, participation is a category that is rooted in the New Testament (Acts 17:28; Col 1:17) and expressed in Christian thought throughout history.

Fifth, creation can be described as sacramental in the sense that it possesses a *mediatorial* function, uniting the creature to the Creator. "The entire world," writes Hans Boersma, "is meant to serve as a sacrament: a material gift from God in and through which we enter the joy of his heavenly presence."[30]

Sixth and finally,[31] creation can be described as sacramental in the sense that it is *typological*, patterned after a greater heavenly reality. According to this use, creation is not simply beautiful and awe-inspiring in isolation; it was designed to communicate something of God's beauty and awesome power. In Marilynne Robinson's novel *Gilead*, the protagonist writes, "It is easy to believe in such moments that water was made primarily for blessing, and only secondarily for growing vegetables or doing the wash. I wish I would have paid more attention to it."[32]

EVALUATING SACRAMENTAL LANGUAGE

In evaluation of sacramental language, I will largely skip—because of space constraints—dedicated treatment of the sacramental as *rite* and *gift*. This is

---

[29]John Milbank, "Alternative Protestantism," in *Radical Orthodoxy and the Reformed Tradition: Creation, Covenant, and Participation*, ed. James K. A. Smith and James Olthuis (Grand Rapids: Baker, 2005), 27 (emphasis original).

[30]Hans Boersma, *Heavenly Participation: The Weaving of a Sacramental Tapestry* (Grand Rapids: Eerdmans, 2011), 9.

[31]Some might argue that these categories are not sufficiently exhaustive. One additional class might be something like "creation is sacramental in the sense that it is *mysterious* and beyond human comprehension, owing to its divine origin and significance." Hans Boersma suggests this when speaking of creation as beyond the full grasp of human comprehension: "The sacramental character of reality was the reason it so often appeared mysterious and beyond human comprehension. . . . We would not go wrong by simply equating mystery and sacrament." Boersma, *Heavenly Participation*, 22. I think this use can be subsumed under the categories of *mediation, participation,* and *typology.* The same can reasonably be said of sacramental language that expresses creation's value; this is better thought of as an implication of creation's sacramental character as conferred by one or more of the aforementioned six senses: "The sacramental perspective reads in nature an importance or inherent value that a purely utilitarian or naturalist point of view cannot discern." John F. Haught, *The Promise of Nature: Ecology and Cosmic Purpose* (Eugene, OR: Wipf & Stock, 1993), 99.

[32]Marilynne Robinson, *Gilead* (New York: Farrar, Straus and Giroux, 2004), 28.

warranted given the extent to which these two senses are both broadly assumed and narrowly debated according to ecclesial tradition. Both senses
inform discussion of a sacramental picture and systematic theology in
general, but they are not essential to understanding the role of the sacramental vis-à-vis re-enchantment.

Sacramental as sacred. Our first question concerns the sacramental as
sacred. Is the world sacramental, if by that we mean it is sacred or holy? Paul
Trebilco thinks it is on the basis of 1 Timothy 4:4-5 (NRSV): "For everything
created by God is good, and nothing is to be rejected, provided it is received
with thanksgiving; for it is *sanctified* by God's word and by prayer." Trebilco
writes, "The Earth community does not consist of disposable matter; rather it
is holy and sacred by virtue of being created and sanctified by God."[33] Dillon
Thornton, a fellow of the Center for Pastor Theologians, describes things similarly in a recent article on the same passage: "The entire cosmos is consecrated."[34]

I have written at length in the *Bulletin of Ecclesial Theology* against this
claim, trying to defend a more narrow understanding of the nature of creation's sanctified, or holy, character.[35] I think inattention to the caveats of
creation's holiness presented in 1 Timothy distorts the meaning of the passage
and warps the larger picture of creation. A more chastened approach recognizes here an incomplete sanctification, conditionally mediated by God's
people, yet still powerfully suggestive of a future, fully consecrated cosmos.
A major element of my argument is the observation that always in the Pastoral Epistles and often in the Pauline corpus, "word of God" means "the
gospel," not the word spoken at creation in Genesis. These textual factors
suggest that only believers of the gospel—hearers of the divine word—partake
of something consecrated, as they are the only ones prepared to offer true
thanks by faith. Rather than arguing for a universally consecrated creation,
Paul thus appears to be making a point in the opposite direction: those sanctified by God, through their grateful prayer of thanksgiving, sanctify in turn—
that is, set apart—that which God has created. I think of this as a current

---

[33]Paul Trebilco, "The Goodness and Holiness of the Earth and the Whole Creation (1 Timothy 4.1-5)," in *Readings from the Perspective of Earth*, ed. Norman C. Habel (Sheffield: Sheffield Academic Press, 2000), 217-18.

[34]Dillon Thornton, "Consecrated Creation: First Timothy 4:1-5 as an Underused Remedy for Cosmological Dualism Prevalent in the Church," *Bulletin of Ecclesial Theology* 4, no. 1 (2017): 21.

[35]Jeremy Mann, "A Consecrated Cosmos? First Timothy 4:1-5 in Exegetical and Theological Per-spective," *Bulletin of Ecclesial Theology* 4, no. 2 (2017): 79-88.

ministry of the church, the kingdom of priests that brings together all the goods of creation and properly orients them toward their Maker.

An eschatological point is also relevant. Prophetic passages such as Zechariah 14:20 depict the expansion of consecrated space into all domains of life: "On that day there shall be inscribed on the bells of the horses, 'Holy to the LORD.' And the cooking pots in the house of the LORD shall be as holy as the bowls in front of the altar" (NRSV). Ezekiel's vision of the Jerusalem to come centers on a central temple where God dwells. This idea is underscored in the final apocalyptic vision of Scripture, where in Revelation 21 an impressionistic picture of the restoration of all things is painted, with Jerusalem as an eternal holy of holies. "The whole city is sanctified through the presence of God and the Lamb; every inhabitant is a 'high priest' with unrestricted access to the inner sanctuary."[36] To universalize creation's holiness presently leaves no room for increased consecration at the eschaton.

If this articulation of the mediatorial role of God's people in consecrating all things to their Maker is correct, we have scriptural and theological reasons for not reading the holiness of creation as presently universal. There is also prudence in avoiding statements that downplay the importance of Spirit-dependent consecration. An example of this is Henri de Lubac's statement that "a man is religious to the very degree that he recognizes everywhere that he lives in a sacred atmosphere."[37] Here recognition *could* be thoughtfully expanded to include proper thanks to God and consecration to his service. It could just as easily be read, however, to imply that a sacred atmosphere permeates all present reality whether it is recognized or not. A similar statement, but perhaps more carefully worded, is found in Wendell Berry: "To live we must daily break the body and shed the blood of Creation. *When we do this knowingly*, lovingly, skillfully, reverently, it is a sacrament."[38]

Sacramental as divine participation. How do we judge the sacramental as a statement about the mingling of creation and God's being? This finds strongest expression in John Milbank, the key figure of Radical Orthodoxy.

---

[36]Dean Flemming, "'On Earth as It Is in Heaven': Holiness and the People of God in Revelation," in *Holiness and Ecclesiology in the New Testament*, ed. Kent E. Brower and Andy Johnson (Grand Rapids: Eerdmans, 2006), 355.

[37]Henri de Lubac, "Internal Causes of the Weakening and Disappearance of the Sense of the Sacred," in *Theology in History* (San Francisco: Ignatius, 1996), 231.

[38]Wendell Berry, *The Gift of Good Land* (San Francisco: North Point, 1981), 281 (emphasis mine).

James K. A. Smith summarizes the core concern of Radical Orthodoxy with four words: "There is no secular."[39] As Milbank, Graham Ward, and Catherine Pickstock see it, this renunciation establishes the critical battle line against a grand campaign threatening the church. The irony of secularism, according to Milbank, is that while it purports to celebrate the only thing it says actually exists—the material world—it actually collapses on itself, given its lack of any larger support. Secularism's rejection of transcendental language leaves only the most impoverished vocabulary. The earth and its culture is at best cosmic flotsam and human litter. Embodied life, self-expression, sexuality, aesthetic experience, and human political economy are all meaningless. The more honest assessment, according to Milbank, is that secularism results in total nihilism.

Against any notion of secularity, a "given" natural, that is, an understanding of being without immediate reference to God, Radical Orthodoxy asserts participation: "the notion of the cosmos as *sharing in*, displaying to a certain degree, the divine essence."[40] The genealogy of the term is significant: "'participation' as developed by Plato and reworked by Christianity, because any alternative configuration perforce reserves a territory independent of God."[41]

According to Milbank, sharing in the divine essence admits of degrees. While criticizing Luther's approach to metaphysics, he writes, "The univocalist-nominalist metaphysic will not allow for a sharing that falls between identity and difference and only imitates by sharing, but equally only shares by imitating."[42] In this construction, the cosmos shares the divine essence, but it is not identical to it. For Milbank, the underlying problem for all species of Reformed theology is dependence on late-medieval philosophy, particularly univocity. By accepting a framework in which all being shares the same single plane of existence, the created is no longer able to nest inside the larger category of divine being, sharing aspects of its essence. This separates created being into its own category, violating the central dogma of Radical Orthodoxy. It also makes it difficult to imagine what sharing in the divine essence—while also remaining distinct—might look like.

---

[39]James K. A. Smith, *Introducing Radical Orthodoxy: Mapping a Post-Secular Theology* (Grand Rapids: Baker Academic, 2004), 88.

[40]Milbank, "Alternative Protestantism," 27 (emphasis mine).

[41]Milbank, Ward, and Pickstock, "Suspending the Material," 3.

[42]Milbank, "Alternative Protestantism," 28.

We must hasten to recognize, however, that participation is not a foreign idea to Christian theology. Traditionally, participation is used to describe the regenerate believer's union with Christ (this is the favored use by Protestants) or the sacraments' divine substance (this is more favored by Roman Catholics). Milbank intends to apply the term more broadly, taking up the preexisting usage but also expanding it "in a new way." He goes on: "Traditional *methexis* concerned a sharing of being and knowledge in the Divine. . . . I have always tried to suggest that participation can be extended also to language, history and culture: the whole realm of human making."[43]

On the basis of Radical Orthodoxy's description of all created things sharing in the divine essence (yet not identical to it), Milbank does not just celebrate its goodness; he takes it as essential for any ascension to heaven: "When we contingently but authentically make things and reshape ourselves through time, we are not estranged from the eternal, but enter further into its recesses by what for us is *the only possible route*."[44] Here we see a particularly clear example of what Milbank and Radical Orthodoxy are occasionally criticized for—a valorization of all domains of human existence, with somewhat sporadic attention to the traditional aspects of Christian faith proscribed by Scripture.

Whereas most Protestant theologians maintain that, while God is present to all creation and grounds its being, "participation" is reserved for restored sons and daughters in Christ (and even then with no small amount of hand-wringing about how to maintain the spiritual inflection of the term), Radical Orthodoxy expands participation to all creation. Milbank and others in the Radical Orthodoxy movement would hardly equate God's presence in all things with God's incarnation in Christ, but the conceptual difference awaits clearer explanation.

A helpful contrast case is found in Thomas Aquinas, who carefully describes three senses of God's presence in creation. God is present "by his power inasmuch as all things are subject to his power," by his knowledge inasmuch "as all things are bare and open to his eyes," and "by his essence, inasmuch as

---

[43]Milbank, *Being Reconciled*, 4.
[44]Milbank, *Being Reconciled*, ix.

he is present to all things as the cause of their being."[45] To illustrate these three senses, Thomas offers examples:

> A king, for example, is said to be present in the whole kingdom by his power, although he is not everywhere present. Again, a thing is said to be by its presence in other things which are subject to its inspection; as things in a house are said to be present to anyone, who nevertheless may not be in substance in every part of the house. Lastly a thing is said to be substantially or essentially in that place in which its substance is.[46]

Later in his work Thomas amends these somewhat to the headings of "divine government," "divine science," and "creation."[47] On the third heading Aquinas writes:

> God is in all things; not, indeed, as part of their essence, nor as an accident, but as an agent is present to that upon which it works. . . . Now since God is very being by His own essence, created being must be His proper effect; as to ignite is the proper effect of fire. Now God causes this effect in things not only when they first begin to be, but as long as they are preserved in being. . . . Therefore as long as a thing has being, God must be present to it, according to its mode of being. But being is innermost in each thing and most fundamentally inherent in all things since it is formal in respect of everything found in a thing, as was shown above. Hence it must be that God is in all things, and innermostly.[48]

In the preliminary objections to the question "Is God present in all things?" Thomas considers the example of demons. Given that there is no fellowship between light and darkness (2 Cor 6:14), surely God cannot be present "innermostly" in a demon. Thomas maintains that God is present, but also present is "the deforming of sin which is not from Him; therefore it is not to be absolutely conceded that God is in the demons, except with the addition, 'inasmuch as they are beings.'"[49] Thus, the positive existence, the being-ness of all created things, is accompanied by the divine presence,

---

[45]Thomas Aquinas, *The Summa Theologica* (New York: Benziger Bros., 1947), I, Q. 8, A. 3.

[46]Aquinas, *Summa Theologica* I, Q. 8, A. 3.

[47]Thomas Aquinas, *On Creation: Quaestiones Disputatae de Potentia Dei, Q. 3* (Washington, DC: Catholic University of America Press, 2011), 164.

[48]Aquinas, *Summa Theologica* I, Q. 8, A. 1.

[49]Aquinas, *Summa Theologica* I, Q. 8, A. 1.

and not just by means of knowledge or power but by ontological grounding. Against Milbank, however, this grounding does not involve the *sharing* of the divine essence, despite the fact that, like Milbank, Aquinas has a robust account of *analogia entis*.[50]

Beyond the blurred boundary between Creator and creature, it is unclear in Radical Orthodoxy how the rupture of creation's peace by sin and the ensuing curse limit celebration of creation's goodness. This omission (despite writing a book with the subtitle *Ontology and Pardon*) perhaps explains Milbank's ambivalence about a more traditional emphasis on Christ's mediatorial office. And while it is hazardous to sharply divide the sacred from the secular, describing the whole created world as participating in God's being obscures the ontological uniqueness of God, "as different from flesh as fire is from water" in Calvin's phrasing.[51]

Sacramental as mediatory and figural. Let us now consider the question of creation's sacramental mediation of God's presence, the fifth category of sacramental language, in conjunction with the final, typological category of sacramental language. On this network of topics, I believe evangelicals are most in the debt of those advocating for a more robust doctrine of creation. Not only is creation beautiful and awe-inspiring, but it was designed to communicate something of God's beauty, his awesome power, his grace, and his perfect peace. Sacramentalists invite the church to more deeply consider creation's mediating function and its figural, illustrative power. All traditions have suffered in some degree from late modernity's profaning influence, but Protestants in particular need to rediscover figural enchantment. The natural wonders, parables about ordinary life on earth, and the evolving motif of God's temples as sites of cosmic harmony are not accidental. Neither are figural comparisons between earthly family members and the Christian's heavenly Father and elder Brother and the church's Bridegroom. Drawing from the Pauline statements that both marriage and the feeding of oxen are dim reflections of a brighter, truer light, a typological reading of creation seeks to relate all that exists to God's higher purposes.

---

[50]The claim, most associated with Thomas Aquinas, that there exists some analogy or correspondence between God and his creation, such that human language can communicate about God despite the infinite difference between the Creator and creation.

[51]John Calvin, *Commentary on John's Gospel* (Grand Rapids: Baker, 1989), 134.

Aquinas reasoned that, as the one who can create out of nothing, God is able to signify things with words like humans can, but he can also signify one thing with another thing—the true referent of the bronze snake on the pole in Numbers 21, hung that all who look on it might be healed, is the bloody God of the New Testament, hung on a cross for a similar yet more expansive purpose. In a diminished way, the Christian visual artist or musician or chef serves the church by exploring the ways the physical world, the "things below," help us better understand the spiritual world, "the things above." This, I believe, also has rich implications for pastors and congregational life. What if kneeling is not just a cultural or personal construct but a way in which our whole person can be lowered and made vulnerable before God? Or what if one of the purposes of wine in communion is to induce some measure of bountiful ecstasy, such that grape juice is actually not a fitting replacement?

Hans Boersma's work is helpful in this regard. One way Boersma avoids the more serious concerns facing Radical Orthodoxy is by approaching the *manner* of God's self-disclosure in creation with less confidence. After all, mystery is an essential element of the sacramentalist's whole approach. When he feels compelled to give some sort of definition, Boersma offers a passage from C. S. Lewis:

> The relationship between speech and writing is one of symbolism. The written characters exist solely for the eye, the spoken words solely for the ear. There is complete discontinuity between them. They are not like one another, nor does the one cause the other to be. . . . Pictures are part of the visible world themselves and represent it only by being part of it. Their visibility has the same source as its. The suns and lamps in pictures seem to shine only because real suns or lamps shine on them; that is, they seem to shine a great deal because they really shine a little in reflecting their archetypes. The sunlight in a picture is therefore not related to real sunlight simply as written words are to spoken. It is a sign, but also something more than a sign, because in it the thing signified is really in a certain mode present. If I had to name the relation I should call it not symbolical but sacramental.[52]

---

[52]C. S. Lewis, "Transposition," in *The Weight of Glory* (New York: Harper, 2001), 112; quoted in Hans Boersma, *Heavenly Participation: The Weaving of a Sacramental Tapestry* (Grand Rapids: Eerdmans, 2011), 23.

There are aspects of Boersma's approach, however, that seem to threaten the freedom of God. This was the concern that similarly animated the late-medieval nominalists in their challenge of elements of a sacramental picture. There is a way of over-reading the symbolic function of creation such that we fail to see how occasionally God uses the desert, which is normally a place of desolation, temptation, and judgment in Scripture, as instead a place of refreshment and God's special restorative power. Sometimes we can read the sacramentalists and discern a concerning priority of the material world to such an extent that the lived piety of a poor person who does not appreciate fine art or leisure gardening is somehow distanced from God. In reality, we often see the opposite. While material blessing is a sign of both God's favor and virtue in the book of Proverbs, in the commandments of Deuteronomy, God warns about the likelihood of comfort and ease resulting in forgetfulness, not gratitude.

To capture this complex picture, I propose a metaphor. Two years ago at the annual conference of the Center for Pastor Theologians, Scott Manetsch presented on Calvin's Geneva. Manetsch mentioned that Calvin instituted an aesthetic in worship that was vaguely disembodied. An example of this rejection of the corporeal aspects of worship can be seen in the use of worship spaces of Geneva. After the stained-glass windows of the grand St. Pierre were shattered by the iconoclasts in Geneva's 1533 Protestant upheaval, nothing was installed to replace them. Eventually, forty-four years later, nets were erected to prevent birds from flying into the former Cathedral.[53]

Imagine that Calvin's approach to the windows of St. Pierre is the opposite extreme of the sacramentalists. On this pole, the Christian does not seek in the created world any illustration or evocation of what God is like, choosing instead a kind of unmediated "spiritual" encounter. Instead of creation serving as occasional guide and reinforcement of the work of God, we deny it any contribution. (Note that this extreme was not in fact Calvin's theological position, regardless of his inattention to the actual windows of St. Pierre. While Calvin was skeptical of any true apprehension of God's character through the figure of nature, he was interested in figural interpretation within the church, once the Christian had put on the "spectacles of Scripture.")

---

[53]Scott M. Manetsch, *Calvin's Company of Pastors: Pastoral Care and the Emerging Reformed Church, 1536–1609* (New York: Oxford University Press, 2013), 33.

On the opposite end of the spectrum, imagine a highly sacramental view that completely encloses our picture of God in a regulated, static set of created images, associations, and notions of God's nearness and work, something like the artistic and representative equivalent of *ex opera operato*.

Both extremes have problems. The universe, as interpreted by Scripture, does show us things about God and often serves as a medium of his communion with us. But there is a danger in relying too much on this indirect depiction, or in constraining the activity of God to that which naturally corresponds to his spiritual nature. We need sturdy windows that help orient us to the heavens, but these windows also need to open, allowing the special movement of God's Spirit to refresh our souls in unexpected and personal ways.[54]

This moderate position, one that attends to creation's representation of heavenly realities but draws its conclusions humbly, captures two additional benefits. First, despite its chastened posture toward the effects of sin, it is also not as liable to obscure creation's inherent, God-given goodness. We must not rush to decode a heavenly reality when we encounter the world God has made. Robert Farrar Capon writes, "Things must be met for themselves. To take them only for their meaning is to convert them into gods—to make them too important, and therefore to make *them* unimportant altogether. Idolatry has two faults. It is not only a slur of the true God; it is also an insult to true things."[55]

The second benefit of a moderate view toward creation's typological character is that the various biblical portrayals of creation can find full expressions. In the Psalms, God's people praise him for all creation, for adorning it, for maintaining it, for supplying all that is necessary for life in it. Only hinted at, and this in the midst of a mutinous occupation, is the idea that God himself would make this lowly yet contested arena his own dwelling place or, more properly, fully incorporate such a humble precinct into his own ineffable habitation. Additionally, the eschaton is not rightly conceived of as simple reform and restoration of the original features of creation.

---

[54]A helpful article that takes a more philosophical approach but with the same instinct is Nicholas Wolterstoff, "Sacrament as Action, not Presence," in *Christ: The Sacramental Word*, ed. David Brown and Ann Loades (London: SPCK, 1996), 101-18.

[55]Robert Farrar Capon, *The Supper of the Lamb: A Culinary Reflection* (New York: Modern Library, 2002), 20.

While creation helps us imagine the eternal kingdom that God's people will one day enjoy, we must never forget how beyond our imagination it truly is. Romans 8 speaks of creation groaning in expectation for full redemption. Gerard Manley Hopkins in his poem "Spring and Death" writes of a "subtle web of black"—that is, Death—that wraps "trees and flowers round" and lies "along the grasses green."[56] While there is purpose and order in God's good world, it does not now fully fulfill its intended function of serving as the dwelling place of God and humanity, and even its reflection of heaven's light is clouded by the curse.

GUIDELINES FOR DEEPER DISCUSSION

Let us consider again a visual artist or musician or chef using the word *sacramental*. On one reading, this is an act of defiance. What has been debased, trivialized, ignored, or fragmented must be defended as the good creation of a holy God. This type of righteous indignation is something theologians should celebrate and imitate. On the more positive side, unguarded use of *sacramental* is like a girl who reaches for a crystal bowl at her grandmother's house, not knowing that while she recognizes something beautiful and heart-stirring, there are prudential rules governing the use of such treasures. The girl's grandmother could scold her for being presumptuous or unmannered and shut the china cabinet quickly. But I think a better response would be for her to take the little girl on her knee at the table and show her each piece of crystal, letting her turn it over in her hands and watch the light paint the walls and ceiling. Then the grandmother will say, "There are very special days we want to celebrate. On those days we cook special food, we use our nicest tablecloths and napkins, and we use this crystal. I would like you to enjoy this sometime, but I want us to make the whole day special when we do."

If the word *sacramental* is a special word, church leaders and those that have been graced by God with a theological education should do their best to help more Christians properly appreciate it. They should also supply them with other good words for more ordinary meals—I am happy to see *typological* and *figural* gaining currency after a period of neglect. I agree with Anthony Kenny that *sacramental* ought to be "related to God not as designer

---

[56]Gerard Manley Hopkins, *Selected Poems* (Oxford: Oxford University Press, 2008), 16.

but as redeemer of the world."[57] The unique gifts of the Lord's Supper and baptism are covenant markers for the people of God, occasions for the merciful grace of God to be experienced and proclaimed. While there is a connection between the use of water, wine, and bread in these ceremonies and the participation of all creation in God's holy character, their significant difference must be underscored. I believe reserving *sacramental* as a term related to the two sacraments is a helpful way of signaling linguistically the like-nothing-else quality of salvation. But one can hold this position and still find common cause with a theologian like Hans Boersma. While precision and care are good, there is something more important. Above all, we must open our eyes to the glory of the world, testify of its power, draw others to see it, and then preach of the God who made it, a God whose glory is without peer.

---

[57] Anthony Kenny, *God and Two Poets: Arthur Hugh Clough and Gerard Manley Hopkins* (London: Sidgwick & Jackson, 1988), 118.

6

Irenaeus, the Devil, and the Goodness of Creation

How Irenaeus's Account of the Devil Reshapes the Christian Narrative in a Pro-terrestrial Direction

GERALD HIESTAND

I F OUR HYMNODY AND FUNERAL SERMONS are any indication, contemporary Christian suspicions abound regarding the goodness of the material world. And if not suspicions about the world's goodness, then at least about its *enduring* goodness. As the old gospel hymn states, the world is not our home; we're just passing through. Heaven, it would seem, occupies pride of place in the popular imagination as the final resting place for the people of God. This heaven-bound narrative is the result of latent Platonic and Stoic influence on patristic Christianity.[1] In the Platonic tradition, the heavenly world of the true "forms" is the dwelling place of all things good. Death is release from the prison of the body, so that the soul can leave the material world and rise to the more perfect world of the

---

An earlier version of this essay appeared in the *Bulletin of Ecclesial Theology* 4, no. 1 (2017): 79-95. Used by permission.

[1] The anti-material posture of Platonic and Stoic thought and its unhealthy influence on Christianity is assumed as a theological premise for this paper, rather than defended as a conclusion. For a helpful analysis of this issue, see James K. A. Smith, "Will the Real Plato Please Stand Up? Participation Versus Incarnation," in *Radical Orthodoxy in the Reformed Tradition: Creation, Covenant, and Participation*, ed. James K. A. Smith and James H. Olthuis (Grand Rapids: Baker Academic, 2005), 61-72.

forms.[2] And in the Stoic account, the way to avoid becoming overly preoccupied with the material world is to recognize that fine dishes are nothing
more than the "corpses of dead animals," that wine is merely "grape juice,"
and that sex is nothing more than the "friction of a piece of gut." When
material things seem "most worthy of our approval," we must instead "lay
them naked and see how cheap they are."[3] Don't make much of the material
world, the Stoic logic goes, because it is not worth making much of. It is all
just "water, dust, bones, stench!"[4]

The anti-material, anti-body posture implicit within these accounts (and
expressed by other ancient Greek philosophers in their own variegated
ways) runs counter to the biblical witness regarding the goodness of the
material world and stands in strong contrast to the Bible's vision of bodily
resurrection and the renewal of the material cosmos. But try as we may,
Christian theology has never been able to wholly shake it. Many of the early
Fathers such as Origen, Tertullian, Clement of Alexandria, Gregory of
Nyssa, and Augustine each in his own way shows a commitment to the
basic Platonic and Stoic prioritization of the "spiritual" over the material.[5]
This Platonic and Stoic narrative has steadily pulled Christian eschatology
up and out of the material world into the world of the forms, gods, and

---

[2]This basic account is woven throughout Plato's writings and can be seen most clearly in his famous analogy of the cave. See Plato, *Republic* 12. For Plato on the benefits of death and the evils
of the body, see such passages as *Phaedo* 63-65, 79a-81d; *Timaeus* 81e; *Apology* 40c-42.

[3]Thus the advice from the Stoic emperor Marcus Aurelius in his *Meditations* 6.13, in *Marcus Aurelius: Meditations with Selected Correspondence*, trans. Robin Hard (Oxford: Oxford University
Press, 2011). The same basic disinterested approach can be seen in other Stoic-influenced Roman
statesman-philosophers such as Cicero and Seneca. For Cicero's comments about death as a
blessing, see his *Tusculum Disputations* 1. Along the same vein, see Seneca's *De Consolatione ad
Marciam* 25. Seneca's perspective about the blessing of death is uncomfortably similar to what
one hears at Christian funerals and what one reads in Augustine's response to the death of his
mother in *Confessions* 12.

[4]Marcus Aurelius, *Meditations* 9.36.

[5]Tertullian, Origen, and Clement of Alexandria are especially noteworthy. For Origen's comments on the vanity of the material body, see *De Principiis* 1.7. For Tertullian's comments about
the dangers of female beauty, see *De cultu feminarum* 1.2–2.2; here Tertullian memorably asserts that makeup and the braiding of hair are dark arts taught to the daughters of men by the
fallen angels of Gen 6. Feminine beauty should not be emphasized but "obliterated and con-
cealed by negligence." For a thorough analysis of Clement of Alexandria, see John Behr, *Asceticism and Anthropology in Irenaeus and Clement* (Oxford: Oxford University Press, 2000).
Behr shows how Clement largely accepts the Platonic and Stoic premise that the unseen world
of the heavens is ontologically superior to the material world, and how this in turns leads to
an overdrawn asceticism.

spirits. The problem with the Platonic eschatological narrative, of course, is that it is wrong. Heaven is not the final resting place for the people of God. God has created us from the earth, as earth people. It is no affirmation of our humanity or credit to God's creative power that we treat the material world (from which we are made) as a throwaway husk. As John's eschatological vision in Revelation 21 and 22 makes clear, the destiny of the Christian—both temporal and eternal—is tied up with this world. What God has made is good, indeed *very* good; it was virginal in Adam, and it will be consummated in Christ.

And here we are not just quibbling about eschatological geography. What's at stake is the very nature of Christian hope. As a pastor I have seen the positive way in which a robust knowledge of our terrestrial future serves as a vital resource for anchoring Christian hope. This hope is in turn the basis of Christian obedience; we persevere in obedience to the teachings and person of Christ precisely because we believe that God's promises are true and his reward is sure. No Stoic ethic, this. As the author of Hebrews makes plain, even Jesus' will to obey was based on his confidence in the eschatological "joy set before him" (Heb 12:2). Visions of disembodied spirits dwelling in an angelic celestial city do little to inspire Christian hope and perseverance. Thankfully, our Lord has more terrestrial things in store for us.

For the purposes of this chapter, I take it as axiomatic that the eternal home of God's people is (at least in part)[6] the earth that now spins through our space and time. Much good work has been done to recapture the Bible's pro-terrestrial posture and its eschatological vision of cosmic hope. As N. T. Wright and others have shown, God's ultimate plan for the material world is not its annihilation but its redemption.[7] So I do not here attempt to make a case that has already been made ably elsewhere. Instead

---

[6] I say "at least in part" because the New Testament also makes plain that Christians are raised up with Christ and seated with him in the heavenly places (Eph 2:6), a position that we occupy for eternity. In the eschaton, we do not trade earth for heaven but rather, in Christ, take on heaven as an extension of our home.

[7] This has been a particular emphasis of the biblical theology movement. See N. T. Wright, *Surprised by Hope: Rethinking Heaven, the Resurrection, and the Mission of the Church* (New York: HarperOne, 2008); Greg Beal and Mitchell Kim, *God Dwells Among Us: Expanding Eden to the Ends of the Earth* (Downers Grove, IL: InterVarsity Press, 2014); J. Richard Middleton, *A New Heaven and a New Earth: Reclaiming Biblical Eschatology* (Grand Rapids: Baker Academic, 2014).

I wish to resource this pro-terrestrial narrative by marshaling the assistance of an unlikely ally—the devil. And not just any old devil, but the devil of the early Christian tradition as articulated by the great church father and bishop Irenaeus of Lyons (ca. 130–200).[8] As we will see, Irenaeus's account of the devil offers us a minority report in the Christian tradition that runs in a somewhat divergent direction from the accounts of the devil that emerged after the third century and that now hold sway in contemporary Christian theology. By retrieving Irenaeus's account of the devil, I hope to resource and bolster biblical theology's soteriological accounts that seek to take seriously the eschatological goodness and permanence of the material world.

THE DEVIL IN IRENAEUS SCHOLARSHIP

Irenaeus stands uniquely among the Fathers. He is rightly called the church's first theologian and is certainly the church's earliest extant biblical theologian. His Christology, anthropology, and early trinitarian articulation offer us perhaps the best look into a developing and maturing second-century Christianity. In many respects, his work established the framework for later Christian reflection. As Gustaf Aulen correctly observes, Irenaeus "did more to fix the lines on which Christian thought was to move for centuries after his day" than did any of the other Fathers.[9] And indeed his thought remains fertile soil for contemporary theological reflection and scholarship. As a consequence, scholarly studies abound regarding Irenaeus's views on apostolic succession, recapitulation, anthropology, Christology, Mariology, canonicity, the rule of

---

[8]It is beyond the scope of this essay to justify my claim that there is such a thing as an "early Christian tradition" with respect to the devil and that Irenaeus represents it. This claim is, in part, the subject of my doctoral dissertation, "'Passing Beyond the Angels': The Interconnection Between Irenaeus' Account of the Devil and His Doctrine of Creation" (PhD diss., University of Reading, 2017), 239-72. Whatever the origins of Irenaeus's narrative, it is clear that he has not constructed it "whole cloth." The main lines of Irenaeus's account of the devil can be found in earlier Christian writers. See Ignatius, *To the Romans* 5; *To the Trallians* 4.2; Papias, *Frag.* 11, 24; Justin, *Apologia ii* 5; Tatian, *Ad Graecos* 7; Athenagoras, *Legatio pro Christianis* 10, 24, 25; and Theophilus, *Ad Autolycum* 2.28-29. No single one of these authors mirrors exactly Irenaeus's account of the devil, yet the similarities point toward a common narrative.

[9]Gustaf Aulen, *Christus Victor: An Historical Study of the Three Main Types of the Idea of Atonement* (Eugene, OR: Wipf & Stock, 2003), 17.

faith, atonement, and divinization (to name a few).[10] And most saliently for the theme of the present volume on the doctrine of creation, Irenaeus is well-known for his strongly pro-cosmological stance.[11] His disputation with the Gnostic heresy compelled him to articulate a clear and aggressive affirmation of the goodness and eventual redemption of the material world—an affirmation that stands unparalleled in the early Christian tradition.[12]

But what is of equal relevance for the present occasion, and what has *not* been explored at length, is Irenaeus's account of the devil and the way this account resources his (and potentially our) high terrestrial cosmology. To be sure, one can find many treatments of Irenaeus that touch on his view of the devil.[13] Likewise, there are a number of scholarly treatments of the devil that touch on Irenaeus.[14] But in both instances, Irenaeus's account of the

---

[10]Major treatments of Irenaeus include Matthew Steenberg, *Irenaeus on Creation: The Cosmic Christ and the Saga of Redemption* (Boston: Brill, 2008), and *Of God and Man: Theology as Anthropology from Irenaeus to Athanasius* (New York: T&T Clark, 2009); John Behr, *Asceticism and Anthropology*, and *Irenaeus of Lyons: Identifying Christianity* (Oxford: Oxford University Press, 2013); Gustaf Wingren, *Man and the Incarnation: A Study in the Biblical Theology of Irenaeus* (Eugene, OR: Wipf & Stock, 2004); Antonio Orbe, *Antropología de San Ireneo* (Madrid: Biblioteca de Autores Cristianos, 1969*)*, *Parábolas Evangélicas en San Ireneo*, vols. 1–2 (Madrid: Biblioteca de Autores Cristianos, 1972), and *Espiritualidad de San Ireneo* (Rome: Editrice Pontificia Universita Gregoriana, 1989); Jacques Fantino, *L'homme image de Dieu chez saint Irénée de Lyon* (Paris: Cerf, 1986); Ysabel de Andia, *Homo Vivens: Incorruptibilité et divinization de l'homme selon Irénée de Lyon* (Paris: Études Augustiniennes, 1986); John Lawson, *The Biblical Theology of Irenaeus* (Eugene, OR: Wipf & Stock, 2006), originally published by Epworth Press, 1948; Denis Minns, *Irenaeus: An Introduction* (New York: T&T Clark, 2010); Ian M. MacKenzie, *Irenaeus's Demonstration of the Apostolic Preaching: A Theological Commentary and Translation* (Hampshire, England: Ashgate, 2002); and Eric Osborne, *Irenaeus of Lyons* (Cambridge: Cambridge University Press, 2001).

[11]See esp. Steenberg, *Irenaeus on Creation* and *Of God and Man*; also Behr, *Asceticism and Anthropology*.

[12]Colin Gunton goes even further, stating that Irenaeus's defense of the goodness of the material creation is "without equal in the history of theology." Gunton, *The Triune Creator: A Historical and Systematic Study* (Grand Rapids: Eerdmans, 1988), 62.

[13]Typically, the devil shows up in Irenaeus scholarship as it relates to the broader themes of atonement. Gustaf Aulen's classic work, *Christus Victor*, 16-35, is a standard here. Likewise, see Gustaf Wingren's *Man and the Incarnation*, chap. 11. Wingren shows how the *Christus Victor* framework undergirds the whole of Irenaeus's narrative. See also the brief but helpful comments in Minns, *Irenaeus*, 104-7. Yet in each case, no systematic treatment of the devil is offered.

[14]The seminal scholarly work on the history of the devil is provided by Jeffrey Russell. His four-volume work explores the identity of the devil from ancient times to modernity. Russell's work touches on Irenaeus in the second volume, *Satan: The Early Christian Tradition* (Ithaca, NY: Cornell University Press, 1987), 80-106. The primary lens through which Russell assesses the devil is theodicy.

devil features only as a peripheral topic in a larger argument—most typically in discussions centered on atonement and theodicy. Such neglect lacks imagination. As William James once famously quipped, "The world is all the richer for having a Devil in it, so long as we keep our foot upon his neck."[15] And rich indeed is Irenaeus's world, not least because of his devil.

Into this open space I offer an executive summary of Irenaeus's account of the devil and the way this account shapes and informs his high terrestrial cosmology and eschatology. Like any story, the shape of Irenaeus's narrative is significantly influenced by the identity and aims of the narrative's chief antagonist. And it is at just this point that Irenaeus's account of the devil has unique power to reshape our overly Platonized Christian story in a more biblical and pro-terrestrial direction. We will explore the devil's identity as angelic steward of the material world, his envy of humanity's lordship, his assault on humanity, his fall, and his eventual defeat—all with a view to showing how this narrative pushes Irenaeus's reading of the biblical plot line in a decidedly pro-terrestrial direction. By way of a foil, we begin with a brief retelling of the devil narrative that now reigns in the contemporary imagination.

THE FOIL: JOHN MILTON'S DEVIL

Beginning with Origen[16] and achieving a relatively fixed status by the time of Gregory the Great in the sixth century, Christian teaching on the devil took the form now known to us and popularized by John Milton in his magisterial work, *Paradise Lost*. In this now familiar "Miltonic"[17] narrative, the fall of the devil and his angels occurs in heaven prior to the

---

[15]William James, *The Varieties of Religious Experience* (New York: Longmans, Green, 1902), 50.

[16]The account of Satan's fall takes a new turn with Origen. Origen's Neoplatonic framework—particularly his notion of the preexistence of the soul—is influential at this point. Creation and the body are the result of the fall, and thus the fall must take place prior to creation. On this account, Origen must look beyond the canon for Satan's fall, rather than taking the Genesis account at face value, as does Irenaeus. More on this below.

[17]It is, of course, anachronistic to refer to the whole of this tradition as "Miltonic." But given that Milton's *Paradise Lost* has done more to shape the contemporary English imagination on the devil than has any other work, and given that our primary concern is pastoral and theological (rather than strictly historical), I here use Milton as the spokesman for a tradition that he, more than any other, has expressed in its most mature form. As we will see, the same is the case for Irenaeus, who himself does not invent the early Christian account of the devil but nonetheless is its most mature spokesman.

creation of humanity.[18] Satan's besetting sin is pride. Though one of the great archangels, he is not content with his limited status in relation to the Son and so leads a rebellion against God in an attempt to usurp the Son's dominion in heaven. The coup fails, and the devil and his angels (one-third of all angels) are cast out of heaven. Still determined to strike against God, the devil attempts to avenge this defeat by attacking humanity, God's prized possession. The garden temptation and the fall of humanity ensue.

A number of features of this narrative are notable. First, the fall of the devil and the angels occurs before the creation of humanity. Thus the informed reader of the biblical narrative has already been handed a backstory that necessarily shapes the reading of Genesis 1–3, which in turn influences the way one reads the rest of the canonical narrative. Second, the primary and initial conflict of this Miltonic narrative is between God and the devil; indeed, the initial conflict of the narrative occurs before humanity has even entered the story. Humanity becomes involved in the plot's conflict only as an innocent bystander, a civilian casualty of the already existing warfare between heaven and hell. And third (and most significantly), the spoil of war in the Miltonic account is a celestial one; Satan's pride has driven him to attempt to usurp the Son's heavenly throne. In this account, the earth is simply the battleground where two extraterrestrial forces wage war. The story concludes with a celestial focus: the devil is defeated in his war against God by the divine Son of God, faithful humanity ascends to heaven to replace the angels who have fallen, and humanity lives happily ever after in God's eternal home.

As we will see, this Miltonic narrative of the devil mirrors the same basic plot sequence and climax that we find in Platonism's non-terrestrial narrative. In the Miltonic account of the devil, the redemption of humanity and the earth are not necessary features in the resolution of the larger soteriological narrative. Irenaeus's account, however, runs in a different direction.

---

[18]See Augustine's extended discussion on the timing of the devil's fall in his *De Genesi ad litteram* 11.1-26. Augustine is uncertain about when the angels fell. But he is certain they didn't maintain their original righteousness for any significant length of time, falling soon after they were created. In any case, for Augustine, Satan has already fallen prior to his temptation of Adam and Eve.

IRENAEUS'S DEVIL: PRE-FALL IDENTITY

Irenaeus's comments regarding the devil are scattered liberally throughout his two extant works.[19] For Irenaeus, the devil is "Satan,"[20] "the serpent,"[21] the "rebel,"[22] the "adversary,"[23] the "deceiver,"[24] "the author and originator of sin,"[25] the "neighbor of death,"[26] the "accuser,"[27] the "dragon,"[28] the "enemy of humanity,"[29] and "the apostate."[30]

But the devil was not always so diabolical. Irenaeus, like other early Christian writers, posits a "fall" in which the devil apostatizes and becomes the enemy of God, of the good angels, and of humanity. Irenaeus does not offer us an exhaustive portrait of the devil's pre-fall identity (in keeping with his general anti-speculative reading of Scripture). Yet given the paucity of scriptural information available on the topic, Irenaeus has more to say about the devil's pre-fall identity than we might otherwise expect. Two aspects of Irenaeus's thought are notable. First, for Irenaeus, the devil began as the angelic "steward" of our

---

[19]Irenaeus's two extant works are *Adversus haereses*, his major work against the Gnostics (abbreviated hereafter as *Haer.*), and *Epideixis tou apostolikou kērygmatos*, a short summary of the biblical story line (abbreviated hereafter as *Epid.*). Both works were originally written in Greek but now remain complete only in Latin and Armenian translations, respectively (with extended and smaller fragments of the Greek found in other writers). For the Latin text of *Adversus haereses*, see the relevant volumes in A. Rousseau, ed., *Sources Chrétiennes*. For the Armenian of *Epideixis*, see the 1919 edition of *Patrologia Orientalis*, vol. 12, edited by K. Ter-Mekerttschian and S. G. Wilson. The English translation of *Adversus haereses* used in this essay follows *Ante-Nicene Fathers*, vol. 1. The English translation for *Epideixis* is based on Armitage Robinson's 1920 translation from the Armenian. I have modified these English translations at various points, based on my reading of the Latin, as well as updated the translations for smoother English reading. The translation of *Epid.* 12 and 14 used throughout follows the work of Matthew Steenberg in his essay "Children in Paradise: Adam and Even as 'Infant' in Irenaeus of Lyons," *Journal of Early Christian Studies* 12, no. 1 (Spring 2004): 1-22. Bracketed Armenian transliterations are drawn from Joseph Smith, *St. Irenaeus: Proof of the Apostolic Preaching*, Ancient Christian Writers 16 (New York: Paulist Press, 1952). Bracketed Latin terms are from *Sources Chrétiennes*, vol. 406. Bracketed Greek terms are from Steenberg's translation, representing his best guess as to Irenaeus's original term.

[20]*Epid.* 11, 16; *Haer.* 5.21.2.

[21]*Haer.* 5.21.1, 3.23.1. Irenaeus clearly identifies the devil with the serpent in Gen 3; Satan indwells the snake, and for this act God thereafter punitively associates the devil with the snake (*Epid.* 16; *Haer.* preface to book 4).

[22]*Haer.* 3.8.2.

[23]*Haer.* 5.21.2, 4.24.1.

[24]*Epid.* 11.

[25]*Epid.* 16.

[26]*Haer.* 5.22.2.

[27]*Haer.* 3.17.3.

[28]*Haer.* 2.31.3.

[29]*Haer.* 4.24.1.

[30]*Haer.* 5.24.4; *Epid.* 11.

planet, ordained to govern the affairs of the world on behalf of humanity until such time as humanity "came of age" and could govern the world on its own. And second, it was as an angelic steward that the devil and his angels were destined to be subject to humanity. We explore each aspect below.

The devil began as an angelic steward of the material world. Fundamental to Irenaeus's perspective on the devil is the idea that the devil began as an archangel, a "creature of God, like the other angels,"[31] who was divinely appointed as steward of the earth. For Irenaeus, innumerable angelic hosts occupy the seven heavens; each is assigned to various tasks by the Creator.[32] The lowest heaven (the seventh) is our firmament. It is in this lowest heaven that the archangel-soon-to-be-the-devil and his angels reside. In the opening chapters of *Epideixis*, Irenaeus writes, "In the domain [i.e., the world] were also, with their tasks, the servants [i.e., the angels] of that God who fashioned all, and this domain [i.e., the world] was in the keeping of the steward, who was set over all his fellow servants. Now the servants were angels, but the steward the archangel."[33]

As the narrative of *Epideixis* unfolds, this angelic "steward" is the one who tempts Eve and so becomes the devil.[34] Thus in Irenaeus, the devil stands apart

---

[31]*Haer.* 4.41.1. The idea that the devil began as an angel is not original with Irenaeus. See Justin, *Dialogus cum Tryphone* 79; Tatian, *Ad Graecos* 7; Athenagoras, *Legatio pro Christianis* 24; and Theophilus, *Ad Autolycum* 2.28. Russell notes that this view was fixed in the Christian tradition from Theophilus onward (ca. 170). See Russell, *Satan*, 78. Russell's comment implies there were alternative early Christian perspectives on the devil's origin. However, I am not aware of any ancient Christian writer (here I exclude Gnostic writings) before or after Irenaeus who offers an alternative understanding of the devil's original ontology.

[32]See *Epid.* 9.

[33]*Epid.* 11. This same idea of the devil's stewardship is again mentioned briefly in *Haer.* 5.24.4: "Just as if any one, being an apostate, and seizing in a hostile manner another man's territory, should harass the inhabitants of it, in order that he might claim for himself the glory of a king among those ignorant of his apostasy and robbery; so likewise also the Devil, being one among those angels placed over the spirit of the air [*sic autem et Diabolus, cum sit unus ex angelis his qui super spiritum aeris praepositi sunt*], as the Apostle Paul has declared in his Epistle to the Ephesians." The *spiritum aeris* here is a reference to the lowest level of heaven and identifies Satan and the angels as those who dwell in the firmament and, presumably, from this position in the cosmos, exercise their stewardship over the material world. While Justin and Papias speak of angelic stewardship generally, Athenagoras is the only other extant early Christian writer who assigns this role to the devil specifically. See *Legatio pro Christianis* 24: "So also do we apprehend the existence of other powers, which exercise dominion about matter, and by means of it, and one in particular, which is hostile . . . to the good that is in God, I say, the spirit which is about matter, who was created by God, just as the other angels were created by Him, and entrusted with the control of matter and the forms of matter."

[34]See *Epid.* 16.

from the other angels and archangels insofar as he was once the chief steward of the material world and leader of those angels assigned to care for the earth.[35]

Our understanding of Irenaeus's position here is informed by other early Christian writers who explicitly taught some form of angelic stewardship over the material world. So Papias: "Some of them—obviously meaning those angels that once were holy—he assigned to rule over the orderly arrangements of the earth, and commissioned them to rule well."[36] Likewise Justin: "God, when he had made the whole world . . . committed the care of humanity and of all things under heaven to angels whom he appointed over them."[37] And Athenagoras: "For this is the office of the angels: to exercise providence for God over the things created and ordered by him, so that God may have the universal and general providence of the whole, while the particular parts are provided for by the angels appointed over them."[38] Given these accounts, it is likely that Irenaeus has something similar in mind when he speaks of the angels as serving God by "keeping" the domain of the earth. Exactly what this care consisted of is not certain. In pre-first-century Jewish thought, the angels were said to have dominion over nations and peoples,[39] but Irenaeus seems to be suggesting something different—since for Irenaeus angelic stewardship is in place from the very beginning of creation (and thus prior to nations and peoples). Did this stewardship involve ordering the powers of the natural world—the winds, the snows, the rivers, the oceans, and so on? Or perhaps taking care of the animals?[40] Irenaeus does not tell us.

In some of the Gnostic schemes that Irenaeus was combating, the devil's association with the material world was a black mark on the devil's curriculum vitae insofar as spirits associated with the material world were viewed as less enlightened than those above. In the Valentinian system, the

---

[35]Along these lines, Smith observes that for Irenaeus the "steward" (i.e., the devil) and the "servants" under him (i.e., the angels) appear to be uniquely "subcelestial." See Smith, *Proof*, 150.

[36]Papias, *Frag.* 11, in *The Apostolic Fathers*, ed. Michael Holmes, 3rd ed. (Grand Rapids: Baker Academic, 2007), 749.

[37]Justin, *Apologia ii* 5, my translation.

[38]Athenagoras, *Legatio pro Christianis* 24, in *Fathers of the Second Century: Hermas, Tatian, Athenagoras, Theophilus, and Clement of Alexandria*, trans. R. P. Pratten, *The Ante-Nicene Fathers*, vol. 2, ed. Philip Schaff and Clevand Coxe (Peabody, MA: Hendrickson, 1994), 142.

[39]See, e.g., Dan 10:13, 20, which makes reference to the "prince of Persia" and the "prince of Greece" and to "Michael, your prince" (i.e., Daniel's). Justin's singular comment in *Apologia ii* 5 might point in this direction as well.

[40]Something along this line seems suggested in *Shepherd of Hermas* 1.4.2.

material world represented the wrong side of town and owed its origins to fear, grief, death, passions, and ignorance; it was certainly not a place for a respectable spirit to dwell.[41] Thus in the Valentinian system, the demiurge (i.e., the evil creator god) and the devil are mutually slandered in their association with the material world.

For Irenaeus, the material world is inherently good and serves as a visible witness to God's inherent goodness and wisdom. Thus it would be inappropriate to read the devil's primordial association with the material world as a slur against the devil. Rather, the devil's association with the material world serves in Irenaeus to underscore the devil's uniqueness and highlights the egregious nature of his rebellion. We cannot go so far as to say that Irenaeus viewed the devil as the highest of all the archangels, yet the fact that the devil was assigned to care for humanity and humanity's world is an indication of the high honor the devil held at the time of creation. His original assignment, at any rate, was most illustrious.[42]

The devil was destined to be subject to humanity. Irenaeus next introduces us to what is perhaps the most central aspect of the devil's pre-fall identity: the temporary nature of the devil's stewardship, and his eventual subjection under humanity. According to Irenaeus, the devil's stewardship of the material world was always intended to be for a limited duration. "Therefore, having made the man lord [κύριος] of the earth and of everything that is in it, [God] secretly appointed him as lord over those [angels] who were servants [δοῦλοι] in it."[43]

Here we must pause to comment briefly on Irenaeus's concept of human infancy. For Irenaeus, Adam was created as "lord of the earth and all things in it," but was nonetheless created as an "infant."[44] As such, it was necessary

---

[41]*Haer.* 1.5.4.

[42]MacKenzie rightly cautions against reading a strong hierarchical structure into Irenaeus's cosmology. With respect to Irenaeus's broader cosmological framework, MacKenzie writes, "There is no cosmological speculation . . . neither is there any rumination in questions of angelic hierarchy." MacKenzie, *Irenaeus's Demonstration*, 97.

[43]*Epid.* 12.

[44]Irenaeus, with the exception of Theophilus (and possibly Clement), is the only extant early Christian writer to speak about the infancy of humanity at the time of creation. For this idea in Theophilus, see his *Ad Autolycum* 2.25. See also Clement, *Protrepticus* 11, where Clement refers to Adam as a παιδίον τοῦ Θεοῦ prior to his fall, and then remarks that through the fall he became a grown man, ὁ παῖς ἀνδριζόμενος ἀπειθείᾳ. The reference is suggestive but only passing and therefore difficult to associate with Irenaeus's concept of human infancy. See also Clement's comment in

that he "grow, and so come to perfection"[45] before he would be able to properly exercise this lordship. The idea that the first human pair were created as infants occurs five times in Irenaeus—two times in *Epideixis* and three times in *Adversus haereses.*[46] For Irenaeus, the infancy of the first human pair explains their need for a steward to govern on their behalf. Irenaeus writes,

> Therefore, having made the man lord [κυριος] of the earth and of everything that is in it, [God] secretly appointed him as lord over those [angels] who were servants [δοῦλοι] in it. They [the angels], however, were in their full development, while the lord, that is the man, was very little, for he was an infant, and it was necessary for him to reach full development [*karelut'iwn*] by growing.[47] . . . But the man was a little one, and his discretion still under-developed, wherefore also he was easily misled by the deceiver.[48]

Thus humanity, though created as the heir of the world, was nonetheless not yet in "full development." The steward and his angels were to govern the world until humanity came of age.

Irenaeus also comments here that though Adam's lordship over the world was public knowledge, Adam's lordship over the angels was "secret" (*zanxlabar*).[49] Irenaeus scholar Joseph Smith helpfully remarks, "The 'secrecy' is probably to be explained by the fact that man, though lord by right, and destined to rule in fact, was not yet capable of doing so . . . so that his lordship was not yet made known to his subjects."[50] This reading makes good sense given that Irenaeus immediately follows his comment about the secrecy of Adam's lordship with comments about Adam's infancy and the

---

Stromata 3.17, likewise passing and suggestive. Behr sees a clear connection among Irenaeus, Theophilus, and Clement on this point; see his *Asceticism and Anthropology*, 135, 143-44. The idea is absent in early Jewish and Gnostic writings. See Steenberg, "Children in Paradise," 20-21.

[45] *Epid.* 12.

[46] *Epid.* 12, 14; *Haer.* 3.22.4, 23.5, 4.38.1-2.

[47] Smith remarks, "The [Armenian] word so rendered is *karelut'iwn*, which would mean 'possibility.'" See his *Proof*, 150.

[48] *Epid.* 12. Irenaeus refers to the "steward" and his "fellow-servants." Thus I take Irenaeus to mean that Adam's lordship over the "servants" includes lordship over the "steward." Irenaeus's account of the temptation stands in stark contrast to Tertullian on this point. For Tertullian, humanity was created in power and glory, as mature bearers of the image of God. The devil resorts to subterfuge precisely because of humanity's greater power. See Tertullian, *Adversus Marcionem* 2.8-9.

[49] The Armenian term is used only here in *Epideixis*. Variously translated elsewhere as "in secret," "furtively," "stealthily."

[50] Smith, *Proof*, 150n69.

maturity of the angels. Thus I take Irenaeus to mean that the steward and his angels knew that the man had been made lord of the world, but they did not know that this lordship extended even to them.

Even without direct knowledge of Adam's future lordship over the angels, the steward and his angels knew themselves to be caring for the world on behalf of humanity. Thus Irenaeus introduces the devil into the creation narrative not only as a servant of God but more pointedly as a servant of humanity. The devil, much like the steward of a child-king, is granted only temporary leadership of the earth until such time as the heir can assume the full responsibility of his throne.

The stewardship of the devil and the infancy of humanity thus serves in Irenaeus as the alternate "backstory" that sets the stage for the first major action in Irenaeus's narrative—the devil's envy of humanity and the garden temptation.

The Devil's Envy of Humanity

We now arrive at the crux of Irenaeus's account of the devil. Irenaeus tells us that the devil's fall was due to his envy of Adam and Eve, and that the devil's first sin was not a celestial rebellion against God in heaven but a terrestrial rebellion against humanity on earth. The idea of the devil's envy of humanity occurs five times in Irenaeus;[51] the most significant occurrence is in the early chapters of *Epideixis*. We begin with *Epideixis* 11 to establish the context. Irenaeus writes,

> But man he formed with his own hands, taking from the earth that which was purest and finest, and mingling in measure his own power with the earth. For he traced his own form on the formation,[52] that that which should be seen should be of divine form: for (as) the image of God was man formed and set on the earth. And that he might become living, He breathed on his face the breath of life; that both for the breath and for the formation man should be like unto God. Moreover, he was free and self-controlled, being made by God for this end, that he might rule all those things that were upon the earth. And this great created world, prepared by God before the formation of man, was given to man as his place, containing all things within itself. And there were in this place also with (their) tasks the servants of that God who formed all

---

[51] *Haer.* 3.23.3-5, 4 preface, 4.40.3, 5.24.4; *Epid.* 16.
[52] Robinson notes that the Armenian text here is equivalent to the Latin *plasma* or *plasmatio*.

things; and the steward, who was set over all his fellow servants, received this place. Now the servants were angels, and the steward was the archangel.[53]

Notable here is the way that Irenaeus highlights the creation of humanity. Human beings are made by God's own hands, a combination of the best of the earth and God's divine power. Moreover, the sovereignty structure between humanity and the angels is clearly established; humanity, not the steward, is destined to rule over this "great created world."

Irenaeus then goes on in chapters thirteen to fifteen to briefly discuss the creation of the animals and of Eve, as well as the prohibition regarding the tree of knowledge. Having set the stage with the principal actors, Irenaeus introduces the reader in chapter sixteen to the first action of the drama—the devil's envy of humanity and the garden temptation.

> This commandment the man kept not, but was disobedient to God, being led astray by the angel who, becoming jealous of the man and looking on him with envy[54] because of the great gifts of God which he had given to man, both ruined himself and made the man a sinner, persuading him to disobey the commandment of God.[55] So the angel, becoming by his falsehood the author and originator of sin, himself was struck down, having offended against God, and man he caused to be cast out from Paradise. And, because through the guidance of his disposition he apostatized and departed from God, he was called Satan, according to the Hebrew word; that is, Apostate:[56] but he is also called Slanderer. Now God cursed the serpent which carried and conveyed the Slanderer; and this malediction came on the beast himself and on the angel hidden and concealed in him, even on Satan; and man he put away from his presence, removing him and making him to dwell on the way to Paradise at that time; because Paradise receives not the sinful.[57]

---

[53]*Epid.* 11.

[54]"Looking on him with envy" is from the Armenian *c̕arakneal* and is perhaps more literally "evil-eyeing." Smith suggests *baskai'nwn* (envying, grudging) as the underlying Greek for this term. See Smith's commentary in *Proof,* 153. For more on the "evil eye" and envy in the Christian tradition, see George R. A. Aquaro, *Death by Envy: The Evil Eye and Envy in the Christian Tradition* (Lincoln, NE: iUniverse, 2004).

[55]Note the close parallel with *Haer.* 4.11.3, where the Pharisees are the "envious wicked stewards" who resist Christ as he rides into Jerusalem to assume his kingdom—a kingdom that they were to rule until his coming.

[56]Cf. *Haer.* 5.21.2. So too Justin (from whom Irenaeus likely got this etymologically incorrect idea), *Dialogus cum Tryphone* 103.

[57]*Epid.* 16.

The steward is not content to be the steward. He is envious and jealous of the "great gifts of God which he had given to man." Irenaeus does not specify here (or elsewhere) the exact nature of the "great gifts" that invoke the devil's envy. But given the overall context, this must certainly include humanity's lordship over the earth. (Indeed, this is the only divine gift given to humanity mentioned thus far in *Epideixis*.) Irenaeus may also have in mind humanity's creation in the *imago Dei*, which he has already mentioned in *Epideixis* 11. This too would be connected to humanity's lordship over the world, for it is precisely because humanity bears the image of God (by which Irenaeus means the image of the embodied human Son)[58] that humanity is the rightful lord of the world.

The devil enters Paradise in the form of a serpent and assaults Adam and Eve while they are yet in their infancy. The devil is successful as it relates to overthrowing humanity; he causes humanity to be cast out of Paradise. But ultimately the plan fails. The steward is found out by God.[59] In this act of rebellion, the steward has overstepped his boundaries and has become an apostate. He too is cast out of Paradise. Insofar as he used a serpent to disguise himself, the steward is cursed with a perpetual association with the serpent.[60]

Ultimately, then, for Irenaeus the devil's rebellion is as much a rebellion against humanity's lordship over the material world as it is a rebellion against God. Unlike the devil in the Miltonic narrative, Irenaeus's devil does not assault humanity as a means of rebelling against God. Rather, the devil wrongly supposes that his treachery toward humanity will go unobserved by God (he futilely uses the serpent as a cloak).[61] Irenaeus's devil has no aspirations to take on God; his target is humanity, and the throne he seeks is earth's. This way of framing the devil's initial relationship to humanity emphasizes the enmity between humanity and the devil as the chief conflict of Irenaeus's soteriological plot line. To be sure, Satan is an enemy of God; but as concerns the narrative Irenaeus will tell, the devil is principally an

---

[58]See *Haer.* 5.16.2. See also Minns, *Irenaeus*, 74.

[59]*Haer.* preface, book 4.

[60]*Haer.* preface, book 4. Irenaeus will go on to assign the fall of the world and the birth of sin most fully to the devil, for the devil was in his full development, while Adam and Eve were mere children. Thus Irenaeus interprets the divine cursing of Gen 3 to be directed chiefly at the devil; Adam and Eve are cursed only indirectly via the curse of the ground and childbearing. See *Haer.* 3.23.5.

[61]See *Haer.* 5.26.2. See also *Haer.* preface, book 4.

enemy of humanity, for humanity is the rightful heir of the world—the chief object of the devil's desire.

THE DEVIL'S ENVY AND IRENAEUS'S PRO-TERRESTRIAL COSMOLOGY

The theological implications of this narrative are far-reaching, particularly when set against later Milton-like accounts. According to Milton, the primary conflict in the Christian narrative is between God and Satan; the restoration of the earth and repossession of its throne by humanity are largely inconsequential to the resolution of the Miltonic devil narrative. The Miltonic account of the devil fits well with, and indeed enables, Platonizing accounts of the biblical narrative.

As I have noted, many of the church fathers downplayed the significance of humanity and the material creation. For Origen and Gregory of Nyssa, creation itself is a result of the fall (or in Gregory's case a punishment for an anticipated fall) and thus not central to God's redemptive purposes—at least not central in any kind of telic sense. Salvation is about leaving behind the material world and shedding the material body. What is more, in such accounts the destiny of redeemed humanity is to become like the angels, freed from the confines and limitations of the material world and destined to dwell with God in the heavens.[62] While Platonizing church fathers such as Origen and Gregory of Nyssa are careful to leave a place for the body and creation, the overall effect of their synthesis tends to be dismissive of materiality in ways not faithful to the broad contours of the canon. A Milton-like account of the devil enables and reinforces this basic Platonic narrative inasmuch as it tends to sideline the embodiedness of humanity and the materiality of creation as central features of the soteriological story and does not require the reenthronement of humanity over the earth as a necessary conclusion.

---

[62]For examples of "angelic soteriology" in early Christian writings, see *Shepherd of Hermas* 3.9.25, 27; Tertullian, *Adversus Marcionem* 3.9; *Ad martyras* 3; *De resurrectione carnis* 36, 42; *De anima* 56. Tertullian's idea that we become like angels at the resurrection is not a denial of the resurrection of the body. He affirms the resurrection of the flesh throughout his writings and is more careful elsewhere to insist that we do not actually become angels. See *De resurrectione carnis* 62. But his repeated emphasis that the high point of salvation is to become like the angels pushes his soteriology in a celestial rather than terrestrial direction. See also Clement, *Paedagogus* 2.10; *Stromata* 6.13; 7.10, 12, 14; Origen, *Contra Celsum* 4.29; *Commentarii in Joannis* 2.16; *Commentarii in Matthaei* 12.30; Augustine, *De civitate Dei* 11.15; 12.16, 22; 22.1; Aquinas, *Summa theologica* 1.62.5 and 1.93.3, where Aquinas states that angels, insofar as they are endowed with a higher intellect than humans, are in some ways more in the image of God than humanity; and Anselm, *Cur Deus homo* 1.16-18.

But Irenaeus's account of the devil pushes the Christian narrative away from Platonizing and Stoic tendencies and toward a more properly anthropocentric, terrestrial climax. In Irenaeus's view, the devil's fall occurs *within* Scripture as detailed in Genesis 3.[63] Most significantly, the world is the prize that humanity initially possesses and that the devil desires. The devil wishes to be worshiped as God, not by supplanting God in heaven, but by supplanting Adam on earth. In short, the devil seeks Adam's throne on earth, not Christ's throne in heaven. What's more, in Irenaeus's account, Satan is a successful usurper of Adam's throne, rather than a failed usurper of Christ's. Thus the primary conflict in Irenaeus's narrative is between the devil and humanity, and the lordship of the material earth is the chief spoil of war.

Humanity's loss of the world's throne via the subterfuge of the devil thus sets the stage for the outworking of the soteriological and eschatological narrative that Irenaeus will tell. Not content with a reversal of earth's lordship, God enters the war between the devil and humanity on the side of humanity and, through Christ, the second Adam, reclaims the world's throne on behalf of humanity. Thus the reclamation of Adam's throne from the devil and the restoration of the material world become central to Irenaeus's biblical narrative in a way not seen in the Miltonic account.[64] With Irenaeus, the soteriological narrative *necessarily* climaxes with the defeat of the devil and the terrestrial reenthronement of humanity; Platonic escape from the material

---

[63]Origen is the first to interpret Ezek 28 and Is 14 as references to Satanic pride vis-à-vis God, and even then only tentatively. See *De principiis* 1.5, 8.3. See also Russell, *Satan*, 125-32.

[64]A fundamental question that must remain unaddressed in this chapter is the extent to which Irenaeus's account of the devil maps onto the biblical story line. That Irenaeus's account of the devil forestalls unwarranted Platonic/Stoic tendencies in Christian theology and eschatology does not mean that Irenaeus's account is true. The question bears more investigation than I can supply here, but I offer briefly five reasons for embracing the early Christian account over the later account: (1) the early account of the devil is early; (2) the early account is reasonably unified in the first two centuries; (3) the early account fits better than the later account with the overall arc of the canonical plot line, which ends in a restoration of the material world and the reenthronement of humanity on the world's throne; (4) the early account postulates that a creature under the authority of Adam was responsible for Adam's downfall; this follows the same basic framework of the Genesis account, which likewise suggests that a creature under Adam's authority was instrumental in humanity's downfall; and (5) the early account is less speculative than the later account, insofar as it places the fall of the devil within the canonical plot line and does not rely on a speculative, precanonical celestial fall. In the end, both the early and later accounts of the devil are speculative to varying degrees; the Bible does not offer us a complete picture of the devil's pre-fall identity and motivations and post-fall activity. But Irenaeus's account is less speculative than the later account and thus to be preferred.

world, or Stoic indifference, is thereby rendered—from the outset—an inadequate consummation to the Christian narrative.

In a remarkably biblical way, Irenaeus's pro-material account of the devil affirms the goodness of the material world against pagan Greek philosophy while it undercuts the temptation to make an idol of the good world that God has made (the opposite pagan error). In some respects, Irenaeus's strong affirmation of the material world may seem to be a counterintuitive way to combat the idolization of it. We might expect that the surer way forward is to chastise the creation, following the route of Platonism and the Stoics. Irenaeus is not naive about the dangers of idolatry. But he would have us break free from idolatry not by dismissing God's good creation but rather by giving thanks for it.

> All [things] have been created for the benefit of that human nature which is saved [*pro eo qui salvatur homine factur sunt*]. . . . And therefore the creation is devoted to humanity [*Et propter hoc condition insumitur homini*]; for humanity was not made for its sake, but creation for the sake of humanity. Those nations however, who did not of themselves raise up their eyes unto heaven, nor returned thanks to their Maker [*neque gratias egerunt factori suo*], nor wished to behold the light of truth, but who were like blind mice concealed in the depths of ignorance, the word justly reckons "as waste water from a sink, and as the turning-weight of a balance—in fact, as nothing."[65]

Creation has been made by a good God for the sake of his people. It has been "devoted" to humanity and thus is to be enjoyed by humanity. The problem, Irenaeus tells us, is not that we like these good gifts too much but that we have forgotten to "return thanks to our Maker." Irenaeus here is following the logic of Paul in Romans 1:18-25, where Paul tells us that the things that are made reveal God's "eternal power and divine nature" (Rom 1:20 ESV). For Paul (and Irenaeus), creation has an iconic function: it is a gift that points beyond itself to the Giver. And as with any "icon," creation derives its value and meaning from that to which it points, namely, God. But humanity, rather than viewing creation as an icon—a springboard—leading to a knowledge of God, instead severed the connection between the icon and the Creator. We fixated on the gift and lost sight of the Giver. But how did

---

[65]*Haer.* 5.29.1.

this breakdown occur? The answer is found in Romans 1:21, which serves as the fulcrum of Paul's logic in this passage: "For although they knew God, they did not honor him as God or give thanks to him" (ESV). The problem is not that we didn't recognize the iconic nature of creation but rather that we failed to give thanks for the icons. This lack of gratitude sets in motion the rest of the story of sin that Paul will address in Romans.

To give genuine thanks for the creation is to acknowledge that there is One above and beyond us who has given it. To give thanks for the world and our bodies necessarily compels us to acknowledge that the Lord *is*, and that he is *good*, and that he *gives*. It reminds us that we ourselves are not the good God but that we stand in a posture of humility and need—recipients of grace. Thankfulness rightly orders our self-understanding with respect to the creation of which we are a part and to the God who made and gave it to us. This is why a refusal to give thanks to God for the good world he has given and a refusal to acknowledge the iconic nature of creation go hand in hand. At its core, thankfulness establishes the relationship between the gift and the Giver. It is impossible to give genuine thanks to God for the good things of the world while idolizing these things at the same time.

The basic contours of Irenaeus's devil narrative do not encourage us to view the material world as a throwaway husk, a ladder to be climbed and then kicked away once we've reached the angelic top. Irenaeus's pro-material account of the devil reminds us, right at the beginning of the Christian soteriological narrative, that creation is a good gift, given to us by a good Creator. It encourages us to view the materiality of creation as a great blessing that God has given to humanity and to view our world as the crown jewel of all the worlds that God has made. Irenaeus's devil narrative tells us that our home is a prize so rare that one of the high archangels of heaven has waged war to possess it. It reminds us that Christ has come not only to save our souls but also to save our home—indeed, to save *his* home insofar as he too is now forever the embodied Son of Man.

7

Wendell Berry and the Materiality of Creation

STEPHEN WITMER

In 1969 an unknown thirty-five-year-old farmer published his first collection of essays, *The Long-Legged House*. Wendell Berry's path to farming was an uncommon one, consisting of a graduate degree, a fellowship at Wallace Stegner's Stanford writing seminar, a Guggenheim fellowship in France and Italy, and a teaching post at New York University. He left NYU in order to teach English at the University of Kentucky, closer to his roots in Henry County, Kentucky. But weekend visits to his farm soon led to a bigger decision—in 1965 he moved with his young family to Lane's Landing, a small farm near Port Royal, the area his family had farmed for generations. In the fifty years since, he has farmed while producing a steady flow of poems, novels, and essays. His body of work and wide influence was recognized in 2010, when President Obama awarded him the National Humanities Medal.

Wendell Berry has never produced a systematically expressed theology of creation. However, he writes frequently and insightfully about the creation, always pushing toward practical concerns. Many of his writings address the question of what is necessary for humans to make good use of the world. Moreover, his answers are grounded in a Judeo-Christian worldview, and the contours of a creation-fall-redemption narrative can be traced throughout his work. Though never fully satisfied with his inherited Protestantism and what he perceives as its dualistic piety, Berry identifies himself (sometimes reluctantly) as a Christian.[1] He calls not for a rejection of Christianity but for a renewal of Christian thought and practice.[2] In essays such as "Christianity and the Survival of Creation" and "The Gift of Good Land," he seeks explicitly to ground this renewal in a reading of the Bible.

Berry's work is rooted in the belief that God is the Creator of the world. He made the creation good. The natural world is bigger than humans, who occupy only a small part of it for short periods of time. Berry demonstrates this in his novels. As Hannah Coulter walks beside the Shade Branch stream, she delights in the sound of rushing water. "You walk up and stand beside it, loving it, and you know it doesn't care whether you love it or not. The stream and the woods don't care if you love them. The place doesn't care if you love it."[3] When Jack Beechum dies, his fields "break free of his demanding and his praise. He feels them loosen from him and go on."[4] Creation is big and humans are small.

And yet, despite their smallness, humans occupy a singular place in creation. God has given them the world as a gift[5] to be stewarded carefully.[6] Berry critiques the bifurcated view of creation found in much modern thought (including that of some conservationists), according to which the creation is either pristine wilderness or ill-used, ravaged land. Drawing on a biblical understanding of stewardship, Berry interposes between the extremes of unused and misused landscape a middle category of *well-used* landscape. Well-used land is stewarded by "humanity at its best,"[7] and this brings glory to God: "good human work honors God's work."[8]

Equally foundational to Berry's thought is the fallenness of creation.[9] For Andy Catlett's grandmother Dorie, "the lostness of Paradise was the prime

---

[1] P. Travis Kroeker, "Sexuality and the Sacramental Imagination: It All Turns on Affection," in *Wendell Berry: Life and Work*, ed. Jason Peters (Lexington: University of Kentucky Press, 2007), 135.

[2] Wendell Berry, *Sex, Economy, Freedom & Community* (New York: Pantheon, 1993), 96.

[3] Wendell Berry, *Hannah Coulter* (Washington, DC: Shoemaker & Hoard, 2004), 85.

[4] Wendell Berry, *The Memory of Old Jack*, rev. ed. (Washington, DC: Counterpoint, 2001), 146.

[5] Wendell Berry, *The Gift of Good Land: Further Essays Cultural and Agricultural* (New York: North Point, 1982), 270.

[6] Berry, *Sex, Economy, Freedom & Community*, 96-97.

[7] Wendell Berry, *Another Turn of the Crank* (Washington, DC: Counterpoint, 1995), 72.

[8] Berry, *Sex, Economy, Freedom & Community*, 104.

[9] Berry, *Gift of Good Land*, 268.

fact of her world, and she felt it keenly."[10] This reckoning with fallenness is evident on almost every page of the novels and essays, which resist sentimentality and grapple frequently with relational, emotional, spiritual, psychological, and ecological brokenness and dysfunction. Jarrett Coulter is so embittered by the death of his wife that he cannot raise his two sons. Virgil Feltner dies in the war, leaving behind a young wife. Nightlife Hample is mentally disturbed. Some marriages are unhappy; some farms are poorly stewarded and then lost; strip mines scar the earth.

When writing of the kingdom of God, Berry usually focuses on what humans do in the present,[11] rather than developing a future eschatology. Nonetheless, there are hints of an eschatological vision of a perfected, embodied future. At the close of the novel *Remembering*, the amputee Andy Catlett dreams of the future new creation and raises in salute to the people of the new Port William "the restored right hand of his joy."[12] Berry writes in his seminal essay "The Body and the Earth" that the Bible doesn't seek to free the spirit from the world; it says they're meant to be reconciled in harmony. "What else can be meant by the resurrection of the body?"[13]

My purpose in this essay is to explore the materiality of creation in Berry's thought. This recognition of materiality enables us to address the larger question of how humans can make good use of creation. Moreover, it leads to important critiques of both institutional Christianity and elements of modern industry and science.

THE MATERIALITY OF CREATION

What does it mean for humans that we are alive *in bodies*, that we participate in a material world through material bodies?[14] Berry addresses this question in various ways.

---

[10]Wendell Berry, *Andy Catlett* (Berkeley, CA: Counterpoint, 2006), 37.

[11]J. Matthew Bonzo and Michael R. Stevens, *Wendell Berry and the Cultivation of Life: A Reader's Guide* (Grand Rapids: Brazos, 2008), 29.

[12]Wendell Berry, *Remembering*, in *Three Short Novels* (New York: Counterpoint, 2002), 222.

[13]Wendell Berry, "The Body and the Earth," in *The Unsettling of America: Culture and Agriculture* (Berkeley, CA: Counterpoint, 2015), 113.

[14]I don't have space in this chapter to explore the implications of our embodied nature for racial identity, though Berry does address race in his writings. It's significant that Ta-Nehisi Coates, *Between the World and Me* (New York: Spiegel & Grau, 2015), 12, frames the question of how he wishes to live this way: "How do I live free in this black body?"

Bodies. Berry's novels frequently and attentively describe human bodies. Marce Catlett is "lean and hard-fleshed," while Jack Beechum is stoutly built, a draft horse of a man.[15] Dave Coulter's nose was "crooked like a hawk's and his eyes were pale and blue."[16] Physical beauty and skill are celebrated. As a young man, Jack Beechum rides his horse to church, "wearing a black suit, fairly new, so made that he keeps a continuous awareness in his waist and shoulders of the perfection of its fit."[17]

But bodies change, falter, fail. As an old man, Jack Beechum is "mindful of the way the weight of his body is taking him." He props it with a cane, allowing his mind to leave it in daydreams.[18] Bodily changes require accommodation and recalibration. The weight of her baby requires pregnant Hannah Coulter to lean back, counterbalancing her "shifted center."[19] Bodies can be funny. Jayber Crow is long of body, legs, arms, hands, face, and nose. "He is all a morose, downward-hanging length."[20] Nine-year-old Andy Catlett thinks he's too skinny, that his legs are rather "grasshopperly" appendages.[21] Big Ellis "was a man of large girth and small behind, who customarily did whatever he was doing with one hand while holding up his pants with the other."[22]

Human beings invariably experience the world in and through a material body, which itself is part of the material creation. No one relates to the world as mere mind or spirit.[23] As Colin Gunton notes in *The Triune Creator*, "The body is that sample of the material creation by which the human being in-dwells and so is related to the created world as a whole: the part both representing and being the means of relating to that part of the whole in which each particular person is placed."[24] The interaction is complex: our bodies

[15]Berry, *Andy Catlett*, 70.
[16]Wendell Berry, *Nathan Coulter*, in *Three Short Novels* (New York: Counterpoint, 2002), 21.
[17]Berry, *Old Jack*, 33.
[18]Berry, *Old Jack*, 3.
[19]Berry, *Old Jack*, 71.
[20]Berry, *Old Jack*, 46.
[21]Berry, *Andy Catlett*, 43.
[22]Wendell Berry, *That Distant Land: The Collected Stories* (Washington, DC: Shoemaker & Hoard, 2004), 253.
[23]For a fresh angle on the importance of our embodied connection to the world, see the fascinating literature on understanding *animals'* embodied experience of the world; e.g., Charles Foster, *Being a Beast* (London: Profile Books, 2016), and Alexandra Horowitz, *Inside of a Dog: What Dogs See, Smell, and Know* (New York: Scribner, 2009).
[24]Colin Gunton, *The Triune Creator: A Historical and Systematic Study* (Grand Rapids: Eerdmans, 1998), 235.

both shape, and are shaped by, the creation. In Jack Beechum's early days on his farm, it feels to him that the rest of his life lies "unborn in the soil of the old farm." Andy Catlett says of his grandfather that "his very flesh . . . had been shaped by weather, work, and the struggle to keep what he had and what he loved."[25]

All this materiality is to be embraced rather than denied. The goal of the Bible is not to free a pure spirit from tainted flesh but to affirm the inescapable mutuality and unity of body and spirit, and teach how they might exist harmoniously.[26] Because our bodies are part of the creation, we must ask "religious questions" such as "What value and respect do we give to our bodies? What uses do we have for them? What relation do we see, if any, between body and mind, or body and soul?"[27] This embodied connection to creation is a recurring theme in Berry's fiction, expressed with particular insight and poignancy in his frequent descriptions of human hands.

Hands. Hands connect us to one another in community, even across generations. In the simple, tender touch of Mat Feltner's hand on his shoulder, Old Jack Beechum feels the touch of his dead nephew, Mat's father, Ben.[28] When Jack dies, Mat is the first to discover his dead body. Mat once again touches his shoulder: "that touch of the hand, that welcome or farewell, by which Ben Feltner was bound to Jack, and Jack to Mat, and Mat to his dead son and to his living grandsons."[29]

Our hands connect us also to the creation, by doing our work in the world. As a man in the prime of his life, Jack Beechum's hands "were not fastidious." They were willing "to do whatever was necessary, to grasp whatever hold was offered, to castrate and slaughter animals, to compel obedience from horse or mule, to cover themselves with whatever filth or dirt or blood his life required."[30] After Jack's death, when the funeral director has worked on his body, Jack no longer looks anything like himself, except, tellingly, for his work-hardened hands, which retain their true character.[31]

---

[25]Berry, *Andy Catlett*, 21.
[26]Berry, *Unsettling*, 113.
[27]Berry, *Unsettling*, 101. See also Berry, *Another Turn*, 91.
[28]Berry, *Old Jack*, 14-15.
[29]Berry, *Old Jack*, 148.
[30]Berry, *Old Jack*, 44.
[31]Berry, *Old Jack*, 154.

In *Remembering*, Berry explores the embodied connection of humans to community and creation by considering the effects of a radical *alteration* of the body. Andy Catlett loses his right hand in a farming accident. "His right hand had been the one with which he reached out to the world and attached himself to it. When he lost his hand he lost his hold. It was as though his hand still clutched all that was dear to him—and was gone."[32] Andy's hand had joined him to his wife in the act of sexual union. It had joined him to creation as he performed the farming tasks necessary to support his family. It had connected him to his community, enabling him to give and receive work from his neighbors. The loss of his hand calls all these connections, and therefore his sense of belonging, and therefore his very identity, into question. He lives in a kind of exile.

Remembering tells the story of Andy's reconnection to the world and his community. His "re-membering" into the membership (i.e., the community) of Port William occurs through remembering,[33] as Andy recalls his own history and that of his community and forebears. Importantly, he recalls the physical touch of members of his community (his father, his grandmother, Nathan Coulter), guiding him back into belonging. There are no easy solutions for Andy. His body is physically altered, and adjustments will be necessary. However, he knows he's connected to a community that is itself connected to creation: "He is caught up again in the old pattern of entrances: of minds into minds, minds into place, places into minds. The pattern limits and complicates him, singling him out in his own flesh."[34] He is *limited* by belonging to this larger reality, because he recognizes he is himself, not any person he might imagine himself to be. He is *complicated* by belonging, because he's forced to deal with his limitations rather than escaping them. He is singled out in his own flesh: he belongs in his body, and his body belongs in his world.

Sex. The core relationship that must be mended in Andy's world, to which he must reconnect (now minus his right hand), is his marriage to Flora. His dismemberment has resulted in disconnection, as he's pulled away from her in confusion and self-pity. In the penultimate chapter of the novel,

---

[32]Berry, *Remembering*, 142.
[33]Peters, *Wendell Berry*, 68.
[34]Berry, *Remembering*, 167.

titled "Bridal," Andy chooses Flora, rather than the disembodied sexual fantasies he has entertained while walking through an airport.

Human sexuality and marriage are recurring themes of Berry's novels and essays, and are themselves important parts of a theology of the body and of creation.[35] We have sexual drives because our bodies are part of nature. Nature is fertile, always seeking to reproduce. Sexual desire is intrinsically wild and primal. Jack Beechum's desires frighten and repulse his wife because of their primitiveness: "His body bent over her in the dark was like a forest at night, full of vast spaces and shadows and the distant outcries of creatures whose names she did not know."[36]

Because sexual desire participates in the rhythm of creation, it sometimes runs wild and untamed. Uncle Andrew is a philanderer. Burley Coulter has a secretive relationship with Kate Helen Branch that leads to the birth of a son he publicly acknowledges only much later. Jack Beechum is drawn into a tragic adulterous affair. In these instances, untamed sexual desire wounds loved ones and destroys relationships.

However, because as embodied beings we inevitably connect to one another and to the world through our bodies, sexual union, as one of the most basic and intimate of bodily connections, can also connect us especially deeply to one another and to creation. "Sexual love is the force that in our bodily life connects us most intimately to the Creation, to the fertility of the world, to farming and the care of animals. It brings us into the dance that holds the community together and joins it to its place."[37]

Healthy, lasting marriages require not the suppression of raw sexual desire but rather its preservation. In "The Body and the Earth," Berry compares the farm and the human sexual body. The well-used landscape of a well-stewarded farm requires ongoing elements of wildness within it for the sake of health: "That is what agricultural fertility *is*: the survival of natural process in the human order. . . . Similarly, the instinctive sexuality within which marriage exists must somehow be made to thrive within marriage."[38]

---

[35]See Stephen Witmer, "Marital Sex Is Creation Care," The Gospel Coalition, July 16, 2016, www .thegospelcoalition.org/article/marital-sex-is-creation-care. Parts of this chapter are drawn from this essay.

[36]Berry, *Old Jack*, 44.

[37]Berry, *Sex, Economy, Freedom & Community*, 133.

[38]Berry, *Unsettling*, 135.

There's a complex interplay here. The bond of marriage is the element of "human order" that both guards the wildness of sexual desire and is itself sustained by it. Marriage exists within the wildness of "instinctive sexuality," and that sexuality thrives best within marriage. In Berry's view, marriage is the fundamental connection "without which nothing holds."[39] It encourages and preserves sexual fidelity, which serves the practical purpose of uniting the larger community. "The forsaking of all others is a keeping of faith, not just with the chosen one, but with the ones forsaken."[40] A healthy, functioning community, composed of healthy, functioning marriages, will encourage, and benefit from, the joyful fulfillment of sexual desire, just as a healthy, functioning farm encourages, and benefits from, the wildness of fertility and natural growth. Sexual love is "the heart of community life."[41] This is a compelling picture of embodied humans stewarding creation together.

When love is reduced to sex and sex is abstracted from community, one result is pornography. Ironically, pornography focuses rapt attention on the fleshly body but is ultimately a *disembodied* thing, sex abstracted from responsibility and relationship. In *Remembering*, Andy Catlett walks through an airport, admiring the attractive women around him. But ironically, the result of these bodies is the *disembodiment* of Andy himself.[42] Rather than considering these women as they actually are or himself as he actually is, he entertains abstract imaginings. This is emblematic of our culture, which heightens sexual desire while simultaneously severing the bonds of community and commitment that can preserve and sustain it. Andy wonders, "Shall we disappear with our longing, dismembered, in the annihilating flame?"[43]

THE GOODNESS OF MATERIALITY

Berry affirms the goodness of the material creation, and of the human body as part of that creation, while simultaneously acknowledging the brokenness of the body and its desires.[44] He finds in the Bible many affirmations of the

---

[39]Berry, *Sex, Economy, Freedom & Community*, 139.
[40]Berry, *Unsettling*, 126.
[41]Berry, *Sex, Economy, Freedom & Community*, 133.
[42]Berry, *Remembering*, 196-97.
[43]Berry, *Remembering*, 197.
[44]For an evangelical attempt to take the body seriously, see Matthew Lee Anderson, *Earthen Vessels: Why Our Bodies Matter to Our Faith* (Minneapolis: Bethany House, 2011).

goodness of matter. In the incarnation, Jesus took on flesh and entered the material world.[45] Jesus healed the physical afflictions of sufferers, one by one.[46] "Again and again the biblical writers write of their pleasure and wonder in the 'manifold' works of God, all keenly observed."[47]

Berry takes institutional Christianity to task for its denial of the goodness of materiality. The Christian tradition has often separated the spirit from the body and therefore from the earth and the goodness of creation.[48] This disdain for the goodness of the present world has been increased by an otherworldly focus on heaven.[49] Jayber Crow, the bachelor barber of Port William, balks at the sermons he hears from young seminary students, all of which have a common theme: "We must lay up treasures in Heaven and not be lured and seduced by this world's pretty and tasty things that do not last but are like the flower that is cut down." These preachers have a "very high opinion of God and a very low opinion of His works."[50] Berry's critique of institutional Christianity accounts for the relative lack of future eschatology in his writings. Rather than focusing on the world to come, he focuses attention on the goodness of the present creation and the importance of properly stewarding it. When he does picture the new creation, it is a material place, inhabited by embodied human beings.[51]

THE MYSTERY OF MATERIALITY

Over against some streams of Christianity, Berry affirms the *goodness* of materiality. But that's not the only front on which he battles. Over against much modern industry and science, Berry affirms the *mystery* of materiality. The creation is not less than material substance, but its materiality does not make it predictable or fully knowable. The material creation remains, ultimately, a mystery and therefore is worthy of reverence and of preservation.

---

[45]Wendell Berry, *Life Is a Miracle: An Essay Against Modern Superstition* (Washington, DC: Counterpoint, 2000), 101.
[46]Berry, *Life Is a Miracle*, 101.
[47]Berry, *Life Is a Miracle*, 102.
[48]Berry, *Gift of Good Land*, 267.
[49]Berry, *Gift of Good Land*, 276.
[50]Wendell Berry, *Jayber Crow* (Washington, DC: Counterpoint, 2000), 160.
[51]Berry, *Remembering*, 221-22.

Berry's 2000 book *Life Is a Miracle: An Essay Against Modern Super-stition*, his critique of *Consilience* (written in 1998 by Harvard scientist Edward O. Wilson), criticizes the mindset in modern industry and science according to which "every mystery is a problem, and every problem can be solved. A mystery can exist only because of human ignorance, and human ignorance is always remediable. The appropriate response is not deference or respect, let alone reverence, but pursuit of 'the answer.'"[52]

Berry concedes that if something unknown is empirically solvable, it falls within the bounds of scientific investigation. But to claim that whatever is unknown is simply "not yet" known is not itself a scientific claim. It rules out (without evidence) the possibility that some things are simply un-knowable, that there is mystery. This critique is central to *Life Is a Miracle*. In fact, the "modern superstition" referred to in the subtitle of the book is the view that "what we take nature to be is what nature is, or that nature is that to which it can be reduced."[53] It is error to believe we know more about creation than we do, and to believe that creation is simply the sum of its component parts. Overoptimism about our knowledge leads to the further superstition that "knowledge is invariably good"[54] and that it will inevitably enable a better use of the creation.

Berry posits a different view of how and to what extent creation can be known. Our fleshly materiality is fundamental to his epistemology. En-fleshed humans and creatures, living within and as part of a material cre-ation, are each unique and can be known (but never completely) only as individuals living within their place.

Berry's first collection of essays, *The Long-Legged House*, contains ac-counts of his return to Kentucky. The long-legged house of the book's title was a two-room cabin built by Berry's great-uncle on a small strip of land beside the Kentucky River, just miles from where generations of Berry's family have lived. Berry spent time there alone as a teenager. Later in life, while spending long days reading and writing there, Berry awoke to the natural life around him. The more he saw, the more he realized this strip of

---

[52]Berry, *Life Is a Miracle*, 27.
[53]Berry, *Another Turn*, 77.
[54]Berry, *Life Is a Miracle*, 144.

riverbank woods was "more unknown than known."[55] And yet the way to achieve his desire to belong fully to the place was to know it deeply. He saw this knowing as an enormous labor—in fact, the patient labor of a lifetime.[56]

> Though it has come slowly and a little at a time, by bits and fragments sometimes weeks apart, I realize after so many years of just being here that my knowledge of the life of this place is rich, my own life part of its richness. And at that I have only made a beginning. Eternal mysteries are here, and temporal ones too. I expect to learn many things before my life is over, and yet to die ignorant. My most inspiring thought is that this place, if I am to live well in it, requires and deserves a lifetime of the most careful attention.[57]

More than three decades later, writing in *Life Is a Miracle*, Berry makes this claim:

> The uniqueness of an individual creature is inherent, not in its physical or behavioral anomalies, but in its *life*. . . . Its life is all that happens to it in its place. Its wholeness is inherent in its life, not in its physiology or biology. This wholeness of creatures and places together is never going to be apparent to an intelligence coldly determined to be empirical or objective. It shows itself to affection and familiarity.[58]

Life is irreducible and unique. Although living things inevitably have a material existence, they cannot be defined or fully understood merely in material terms, as the sum of their parts, or by an analysis of their chemical or biological components. That sort of reductionistic materialism (as opposed to a recognition of materiality) encourages an overly deterministic view of living creatures. Precisely because material things are *living* things, each one has a unique life history.

Berry highlights an unfortunate and common habit of conceptualizing the human body, as well as the world and its creatures, as machines.[59] But machines are predictable and interchangeable, while fleshly bodies are not. Humans make choices, go in unexpected directions, and change over time. No person is exactly the same as another person; no tree exactly the same

---

[55]Wendell Berry, *The Long-Legged House* (Washington, DC: Shoemaker & Hoard, 2004), 149.
[56]Berry, *Long-Legged House*, 150.
[57]Berry, *Long-Legged House*, 168.
[58]Berry, *Life Is a Miracle*, 40.
[59]Berry, *Life Is a Miracle*, 6.

as another tree. Therefore, the only way to ever fully know any living creature is to observe and understand its entire life history. That's why every creature maintains an ultimate inscrutability and mystery. To know something, albeit imperfectly, we must know it intimately, over time: "It shows itself to affection and familiarity."

To illustrate this claim, Berry describes his view out the window of the long-legged house the morning of his writing: patches of melting snow, a swift and muddy river, a blue heron fishing. He writes out of a familiarity and affection borne from years of watching out the window. And though he knows his patch of woods better now than he did decades before, he writes, "I see that the life of this place is always emerging beyond expectation or prediction or typicality, that it is unique, given to the world minute by minute, only once, never to be repeated. And then is when I see that this life is a miracle, absolutely worth having, absolutely worth saving. We are alive within mystery, by miracle. We have more than we can know."[60]

Berry has come to know his place because he *loved* it—not merely as a collection of parts, but as a whole. He points out that not even scientists themselves believe that reducing whole living beings to their categories or component parts is a sufficient manner of knowing. Scientists don't refer to their family members as "a woman," "a man," or "a child." "Affection requires us to break out of the abstractions, the categories, and confront the creature itself in its life in its place."[61] This awareness of the "preciousness of individual lives and places" comes not from science but from our cultural and religious traditions.[62] The heuristic qualities of affection and delight were well-known to the psalmist: "Great are the works of the LORD, studied by all who delight in them" (Ps 111:2 ESV). We observe most closely when we delight most deeply.

Maintaining awe and affection before the mystery of creation is crucial for preserving it. "Things cannot survive as categories but only as individual creatures living uniquely where they live."[63] While acknowledging that there is a "necessary usefulness" in "knowing the parts of a thing and how they

---

[60]Berry, *Life Is a Miracle*, 45.
[61]Berry, *Life Is a Miracle*, 41.
[62]Berry, *Life Is a Miracle*, 42.
[63]Berry, *Life Is a Miracle*, 41.

are joined together,"[64] Berry points to abstraction as the main limitation of reductionistic ways of knowing. There's a paradox here: "Empirical knowledge of the material world gives rise to abstractions such as statistical averages which have no materiality and exist only as ideas." "Between the species and the specimen the creature itself, the individual creature, is lost."[65] When this happens, the value of individual life is diminished. The paradox is ultimately painful: though deeply concerned with the material world, scientific ways of knowing may arrive "at a sort of fundamentalist disdain for material reality."[66]

And this matters a lot, because when the creation is viewed as merely the sum of its raw material, the individual parts and places of creation are seen as interchangeable, and a price tag may be placed on each. In *Remembering*, Andy Catlett visits an agricultural conference devoted to promoting "the abstractions by which things and lives are transformed into money." Industrial agriculture can convert family life and work "to numbers and to somebody else's profit, but the family cannot be seen and its suffering cannot be felt."[67] As J. Matthew Bonzo and Michael Stevens note, in Berry's view, the inability to see things as they truly are always leads to selfishness.[68] When the material creation remains to us ultimately a mystery, we know it as worthy of reverence and therefore preservation.

We'll also be led to worship. In the Bible itself, close observation of the material creation leads to wonder and praise.

> O LORD, how manifold are your works!
>> In wisdom have you made them all;
>> the earth is full of your creatures. . . .
> I will sing to the LORD as long as I live;
>> I will sing praise to my God while I have being. (Ps 104:24, 33 ESV)

Recognizing the materiality of creation, and therefore the unique mystery of every life in every place, increases our desire to observe and therefore our capacity to praise.

---

[64]Berry, *Life Is a Miracle*, 39.
[65]Berry, *Life Is a Miracle*, 39.
[66]Wendell Berry, *Our Only World: Ten Essays* (Berkeley, CA: Counterpoint, 2015), 7.
[67]Berry, *Remembering*, 139.
[68]Bonzo and Stevens, *Wendell Berry*, 27.

Wendell Berry's long insistence on the materiality of creation is important for contemporary Christians as we aspire to faithful lives of discipleship and stewardship amidst a welter of viewpoints that lead to the diminishing of the body, of sex, and of the rest of the material creation. For decades Berry has maintained a dogged focus on what it means to live well in this world: "What I stand for/is what I stand on."[69]

Berry's theology of a material world is an important resource for the church. Engaging with it will help us as followers of Christ to pursue a deeper, richer, more sustainable relationship with the world around us and so draw nearer to its Creator.

---

[69]Wendell Berry, "Below," in *The Collected Poems, 1957–1982* (San Francisco: North Point, 1985), 207.

8

Creation, New Creation, and the So-Called Mission of God

JOHN H. WALTON

I<small>T IS POPULAR TODAY TO SPEAK OF</small> the "mission of God" and the "mission of the church." Theologians describe that mission in a variety of ways, many of which relate to a perceived metanarrative of Scripture. As we consider such proposals, methodology should be our first consideration. What sort of hermeneutic undergirds the idea of a metanarrative? Positing the presence of a metanarrative suggests either (1) that all authors from the beginning of the project have the larger picture in mind and write toward it or (2) that all the books of the canon were redacted at the end of the process in order to shape them to the metanarrative. Perhaps other options might combine some aspects of each of these.

Another way to get at the question is to ask whether the metanarrative is attributed to God, to the authors, or to the redactors. Who is responsible for choosing and shaping it? The answer will determine what hermeneutic is used to identify the metanarrative. In a full-canon approach to metanarrative, the only party that has input in all the parts is God. Early authors cannot influence what later authors write, and they are not involved in selecting books to fit the criteria of the metanarrative. By the time the New Testament authors come along, the Old Testament books have already been shaped into the canon, so no metanarrative of theirs could shape the earlier choices, though they could impose a metanarrative interpretation. It is common to believe that God (through the Holy

Spirit) is the shaper of the metanarrative, working behind the scenes to craft the mosaic or tapestry.

To support such a view is to abandon many commonly accepted herme-neutical controls in favor of a theological premise. If *God* has provided the metanarrative, rather than the authors or the redactors, then the intentions of the human instruments no longer stand as the focus of our textual and literary analysis. Instead, the metanarrative has been recognized as the mys-terious work of God cobbled together in hindsight by postcanonical theo-logians, although, admittedly, they may find their prompts in the views of the latest authors of Scripture who offer big-picture perspectives. This ap-proach is represented by many Christian scholars throughout history who have seen a christological or redemptive metanarrative. God is seen as crafting the metanarrative throughout the Old Testament, though the au-thors or redactors of the Old Testament had little or no knowledge of it. That work, hinted at by Jesus, continues in the New Testament, whose authors are seen as beginning to promulgate a coherent understanding. Nevertheless, it is left to postcanonical theologians to work out the details of a compre-hensive perspective. The resulting perspective is often christological and redemptive in focus. Yet, as much as we respect these theologians, we rec-ognize that they do not carry the same authority as the Bible and they at times disagree with one another. How then can biblical authority be asso-ciated with such interpretations? This redemptive metanarrative is often the foundation for understanding the church's mission or even the mission of God. But if a metanarrative message is considered to drive the main au-thoritative message of Scripture, what exactly is the locus of that authority?

One example of the redemptive metanarrative relative to the mission of the church is developed by G. K. Beale in *The Temple and the Church's Mission*. He identifies the mission of the church in relationship to its identity as the temple where God's presence is manifest. But as time goes on and theology progresses, Beale identifies the true temple as Christ, and the mission of the church is to advance the presence of God by living for Christ and speaking his word, with the result that unbelievers would accept Christ and Satan would be defeated.[1] Another familiar example that focuses on the mission of

---

[1]G. K. Beale, *The Temple and the Church's Mission* (Downers Grove, IL: InterVarsity Press, 2004), 397.

God rather than the mission of the church is represented by Christopher J. H. Wright in *The Mission of God: Unlocking the Bible's Grand Narrative*. In his view, the mission of God is to restore creation through a redeemed community of people that reflects the character of Yahweh.[2]

The theological benefit of this interpretation is self-evident in its affirmation of the coherence of the biblical text across the canon, but at what hermeneutical cost? Can we afford the luxury of this metanarrative and still maintain a consistent hermeneutical methodology that provides appropriate controls? Those who adopt this redemptive-historical position often look to the New Testament for pulling together the pieces and show little concern for the intentions of the Old Testament authors. I have mentioned Beale and Wright specifically because they both give substantial place to the Old Testament.[3] The main dispute I have with those who take a position similar to Wright's (and they are plentiful) concerns the suggestion that Israel's identity is found in their being a "redeemed" community. It is true that Israel was delivered from slavery in Egypt, but the biblical record attests that this is not their primary identity throughout history. Their identity is that they are Yahweh's chosen people in covenant relationship with him. Their redemption is past, not future. At the same time, Wright's focus on restoring creation has some strengths that can be developed further, but that is a separate issue which we will have little opportunity to explore.[4] I find agreement with Beale's position insofar as it is centered on the temple—the presence of God. Even then, however, I would not be as inclined to associate that as strongly with redemption as he is.

Michael Goheen affirms Wright's statement that "God's mission involves God's people living in God's way in the sight of the nations."[5] He elaborates, saying,

---

[2]For example, see Christopher J. H. Wright, *The Mission of God: Unlocking the Bible's Grand Narrative* (Downers Grove, IL: InterVarsity Press, 2006), 22-23. See his summary in Christopher J. H. Wright, "Mission and Old Testament Interpretation," in *Hearing the Old Testament*, ed. Craig G. Bartholomew and David J. H. Beldman (Grand Rapids: Eerdmans, 2012), 180-203.

[3]Similarly, see Michael Goheen, *A Light to the Nations: The Missional Church and the Biblical Story* (Grand Rapids: Baker Academic, 2011).

[4]The importance of the theme of restoring all of creation has been developed by N. T. Wright in numerous works and in J. Richard Middleton, *A New Heaven and a New Earth: Reclaiming Biblical Eschatology* (Grand Rapids: Baker Academic, 2014).

[5]Wright, *Mission of God*, 470.

Thus the nation of Israel was to be a display people, embodying in its com-
munal life God's original creational intention and eschatological goal for humanity. He would come and dwell among them and give them his torah to
direct their corporate life in his way. God's people were to be an attractive sign
before all nations of what God had intended from the beginning, and of the
goal toward which he was moving: the restoration of all creation and human
life from the corruption of sin.[6]

The reflections of these writers have much to commend them, and my
intention is not to engage in point-by-point critique of these important
works. Instead, I would like to pick up some aspects of their work to
shape an alternative approach that, rather than proposing a metanarrative, uses the concept of common themes in a way that is characterized
by hermeneutical rigor and can withstand exegetical scrutiny. At its
worst, a metanarrative approach is formulated by the creative imagination of the interpreter and then shored up with backfill from proof
texts that fall short of representing the perspectives of the authors of the
individual texts. Though I am not suggesting that the protagonists referred to above operate this way, the fact that some do raises the issue of
hermeneutical controls. How would we ever determine that a proposed
metanarrative is true? Constructive theology works in the arena of determining theological truth all the time, and I propose that we relegate
the idea of metanarrative to theologians, while keeping our exegesis focused on what the authors themselves were doing (literarily and theologically)—text in context. The result of such an approach would be to
think in terms of recurrent significant themes rather than metanarrative.
These have been differentiated in the following way:

A metanarrative has a linear structure—sometimes called a narrative arc—
where each plot point builds on those previously established. Normally an
inciting incident (in the salvation history the inciting incident is the fall) instigates a series of tensions (the conflict) leading up to a climax (in the salvation history the climax is the crucifixion) which is followed by a wrapping
up of loose ends leading to the conclusion. In contrast, recurring themes and
motifs have a more centralized pattern; that is, all of the instances of the motif
or theme reference the same thing (sometimes with different emphases or to

---

[6]Goheen, *Light to the Nations*, 25.

different ends) but do not need to reference each other. Two or more authors who are unaware of each other's work can refer to the same motifs or themes, but they cannot contribute intentionally to the same metanarrative.[7]

As a result of this distinction, a text-in-context approach might find more resonance with the themes of creation and presence than with redemption and salvation—the latter pair taking on importance only in later stages of the unfolding of God's plans in the canon. To say this in another way, approaches that find the mission of God or the mission of God's people in redemption, salvation, or restoration ground their understanding in Genesis 3; that is where sin enters, and it is sin, or the fall of creation, that necessitates redemption or restoration. Can we say nothing of God's purposes that may have existed prior to the fall and carried significance regardless of the fall? Can we speak only of God's remedial work? Those who focus on restoration would perhaps respond that they do not do so but focus on the restoration of Eden.

In the view that I will present here, the mission of God is to be found in creation rather than in redemption, and it is not Eden that is restored. Restoration and redemption, for all their theological significance, stand only as secondary aspects of the larger mission of God—establishing his presence among the people whom he has created for relationship. I call this an Immanuel Theology. The recurrent theme that we can easily trace through the canon and that is recognized on the authorial level throughout is that of the presence of God.

From the seven-day account of creation in Genesis 1:1–2:4 we can discern God's interest in ordering the cosmos to receive his presence, which in turn would establish the concept of sacred space. In Genesis 2:1-3 this is not yet developed clearly because the "rest" that the text indicates is conveyed only by the term *šabbat*, connoting only the cessation of the ordering work of creation. In Genesis 2, however, we observe that God has taken up his residence in Eden. It is therefore sacred space, and Adam and Eve are given the task of maintaining this sacred space (Gen 2:15). The connection of residence with rest is established in Exodus 20:8-11, where God's rest on day seven is described by *wayyānaḥ*, which in turn is correlated with his rule

---

[7]John H. Walton and J. Harvey Walton, *Demons and Spirits in Biblical Theology: Reading the Biblical Text in Its Cultural and Literary Context* (Eugene, OR: Wipf & Stock, forthcoming).

from his residence in the temple in Psalm 132:14.[8] From these we deduce that the Old Testament authors present God as having created the cosmos for his dwelling, that he intended to dwell among his people, and that his presence among them was for the purpose of being in relationship with them. Given these parameters, the mission of God takes the shape of presence for relationship from the beginning—prior to the fall and distinct from the need for redemption from something or restoration to something.

Genesis 1 indicates that people were created in God's image to partner with him by continuing the process of bringing God's order to the cosmos.[9] This is evident in their designated function: to subdue and rule. Eden, then, was not a situation of ultimate order to which creation will be restored. Instead, it represented an optimal order, sufficient for a good beginning, from which God would continue to work alongside his image-creatures to establish the ultimate order that we call "new creation." In this way, from the beginning, God's purpose, his mission if you will, is not to be found in redemption and restoration but in presence and relationship. This emphasis is present and recognized in the books of the Old Testament and is intentionally addressed and developed by its authors and redactors.

In this view, Genesis 3 is not understood primarily as the entrance of sin into the world. Furthermore, the Old Testament never looks back to revisit that moment to explore its significance or probe its implications. The prophets never invoke Genesis 3 in any way as they frequently discuss the problem of sin in Israel. When Adam and Eve ate from the tree of wisdom, they were rejecting partnership with God, desiring instead to position themselves as the source and center of order. In that sense they would "become like God," because order is the result of wisdom. The garden was a place of God's order, and since they had desired to find order on their own, they were driven from the presence of God, where his order reigned, to seek out their own order. In this way God granted them what they had desired, but it was to their detriment and, in the process, relationship was thus broken. Genesis 4–11 traces how people worked at

---

[8]The verb translated "rule" in the NIV is the root *yšb*, which has a broad lexical sense but is used contextually in numerous places for sitting on a throne to rule (see Ps 2:4; 7:7; 9:4-11; 22:3; 29:10; 55:19; 80:1 and many more).

[9]Note that some interpreters consider the image of God as primarily relational; Catherine McDowell, *The "Image of God" in Eden: The Creation of Mankind in Genesis 2:5–3:24 in Light of the* mīs pî, pīt pî *and* wpt-r *Rituals of Mesopotamia and Ancient Egypt* (Winona Lake, IN: Eisenbrauns, 2015).

establishing order for themselves, with mixed results. Eventual, overwhelming disorder resulted in the flood and a reordering of the world as creation is literarily recapitulated in Genesis 6–9. Tracking order and disorder is inherent in the rhetorical strategy of Genesis 1–11, as that theme brings coherence to the wide variety of narratives and genealogies found there.[10]

What is not inherent in the rhetorical strategy of Genesis 1–11 is the idea that God has a plan for future redemption and restoration. That element is traditionally located in Genesis 3:15, which is generally considered the launching pad for the salvation metanarrative. This verse is often treated as the keystone for understanding the mission of God. The problem is that neither exegesis of the Old Testament nor the use of this verse in the New Testament supports such an interpretation. The Hebrew of Genesis 3:15 gives no suggestion of an eventual victory; it only asserts an ongoing conflict. The verbs expressing the actions of both parties are the same root, and the verse indicates the mutual exchange of potentially mortal blows (crushed head versus poisonous bite). No victor is announced. In the New Testament, whatever may be said about the victory of Jesus over sin, death, or the forces of evil, Genesis 3:15 is never picked up to give voice to that. The only possible reference to Genesis 3:15 is in the final greetings of Paul to the church at Rome (Rom 16:20), where he prays that Satan may be crushed under the feet of the *church*, not Christ.[11] It is therefore neither exegetically nor intertextually substantiated to see Genesis 3:15 as launching a metanarrative.[12]

To return to the rhetorical strategy of Genesis 1–11, the climax of this section is found in the Tower of Babel narrative where people attempt to reestablish the presence of God.[13] This focus of the narrative has not been traditionally recognized because the identity of the tower as a ziggurat and the purpose of ziggurats had been lost as early as the Hellenistic period.

---

[10]Developed and articulated in Tremper Longman III and John H. Walton, *The Lost World of the Flood: Mythology, Theology, and the Deluge Debate* (Downers Grove, IL: IVP Academic, 2018), 112-21.

[11]Paul does not pick up the verb "crushed" from the LXX translation of Gen 3:15, and he speaks of Satan being placed under their feet, a general idiom used throughout the Old Testament and the ancient Near East that may have nothing to do with Gen 3:15. For more detail see Walton and Walton, *Demons and Spirits*.

[12]Genesis 3:15 is in the context of the first *toledot* (account) of Genesis (2:4–4:26), and a case can be made that its implications pertain specifically to the following sections within that *toledot*. For details of how that can be substantiated, see Walton and Walton, *Demons and Spirits*. Note that in Genesis, none of the elements within a *toledot* implies or refers to later *toledoths*.

[13]Described in detail in Longman and Walton, *Lost World of the Flood*, 129-42.

ancient Near Eastern texts combined with archaeological excavations have demonstrated that ziggurats were built to facilitate God's descent to his temple to take up his residence there and be worshiped. The builders were thus trying to regain the presence of God. Their initiative was rejected, however, because it was motivated by the desire to make their own name great. This continues to reflect the mentality adopted by Adam and Eve in Eden. The proper restoration of the presence of God would exalt his name. Instead, people wanted to exalt their own name; they still want order (brought by God's presence) to be based on them. Consequently, we can see that there is an inclusio of Genesis 1–3 with Genesis 11, and that the internal rhetorical strategy given to Genesis 1–11 by its author or redactors concerns the issue of order as it is related to the presence of God.

This focus continues when we turn to Genesis 12 and the covenant with Abram. Those who adopt a redemptive metanarrative impose on the covenant a redemptive purpose. This begins with the understanding that when Abram is told that through him all the nations of the earth would be blessed, the covenant refers to Jesus—clearly an idea that neither Abram nor any of his descendants in the Old Testament would have understood.

The alternative that I would suggest connects directly to what has been going on in Genesis 1–11 and is thus defensible as inherent in the author's or redactors' intentions. In contrast to the failed human initiative to reestablish the presence of God reflected in Genesis 11, the covenant (literally right on its heels) represents God's counterinitiative to reestablish his presence. We should note that nothing at this stage hints at redeeming or reconciling.[14] He does so by initiating a relationship, his overall mission, through the covenant. This relationship is the prelude to God's intended goal to reestablish his presence on earth. He is going to dwell in the midst of his people, and thus, through them, all the nations of the world will be blessed—by God's presence on earth. This is accomplished initially in the tabernacle, then in the temple. The covenant is made with the goal of God's presence being established. Leviticus 26:11-13 articulates this clearly: "I will put my dwelling place among you, and I will not abhor you. I will walk among you and be your God, and you will be my people. I am the LORD your God, who brought you out of Egypt so that

---

[14]Some might object that there is no hint of presence either, but, as will be seen, the covenant does eventuate in presence in the Pentateuchal context—it does not eventuate in redemption from sin.

you would no longer be slaves to the Egyptians; I broke the bars of your yoke and enabled you to walk with heads held high." Here relationship and presence are foremost as the focus of the covenant. Reference is made to God's delivering them from Egypt but not to future redemption or restoration. The fact that this theme, expressed by these words, is reiterated throughout the remainder of Scripture[15] demonstrates that the authors and redactors of both testaments were aware of it and developed the theme in a variety of ways.

Throughout Exodus, the text is moving toward the reestablishment of the presence of God, from the burning bush, the plagues, and the pillar of cloud, to God's descent on Mount Sinai and then, the climax, taking up his residence in the tabernacle. The torah is not the main focus at Sinai—it is providing instruction for God's people to live in God's presence. It represents God's order for his people Israel and so establishes covenant order in contrast to the order that people had been seeking to establish for themselves. God's presence is the source and center of this order, and the relationship that God has with Israel in the covenant is both the premise and the result. The torah (along with its rituals) provides the instrument for maintaining order in the covenant relationship, but it offers no anticipation of future redemption or the restoration of Eden in particular or creation in general.

God offered Abram a land in connection with the covenant. This is not arbitrary or incidental. God is going to reestablish his presence, and that calls for a land in which he will dwell. Old Testament theology is clear enough that it is Yahweh's land. The enactment of *ḥerem* is not inherently an act of destruction but an act of eminent domain that is clearing the land for Yahweh's presence. Israel receives tenancy of the land as his covenant people who will be the hosts and caretakers of his presence. The land comes to Israel as an establishment of order essential to the presence of God, not as an act of conquest to destroy other peoples. The repeated importance of driving out the Canaanites is that their presence would be detrimental to Israel's identity in Yahweh. To maintain covenant order in the presence of Yahweh, Canaanite influence and identity must be eliminated.[16] Still no indication of redemption or restoration is evident.

---

[15]Among the more prominent occurrences, note Jer 30:22; Ezek 34:30; 2 Cor 6:16; Rev 21:3.

[16]For development of this interpretation, see John H. Walton and J. Harvey Walton, *The Lost World of the Israelite Conquest: Covenant, Retribution, and the Fate of the Canaanites* (Downers Grove, IL: IVP Academic, 2017).

During the period of David and Solomon, the presence of God is transferred to the temple, and Solomon's dedicatory prayer indicates that the same elements continue to be recognized as primary. He refers to the covenant and torah (1 Kings 8:56-61), all in connection to God's presence being with them (v. 57). This status quo remains throughout the monarchy period until the seventh-century classical prophets who begin to pronounce the judgment that God's presence is going to be lost and the temple destroyed. In this context the conversation finally turns to restoration, certainly of the covenant and God's presence, but also of creation in general. At this stage it becomes one of the objectives that will contribute to God's overall goal of dwelling among his people and being in relationship with them (e.g., Ezek 34–36).

All the passages that we have examined in the Old Testament reflect this overall theme of God's presence (eventually located in the temple where he has chosen to place his name) and relationship between God and his people (initiated through the covenant and outlined by the torah). These themes are purposefully and repeatedly addressed in a variety of books of the Old Testament and are pivotal in the native Israelite theology expressed in the Old Testament.[17]

This pervasive Old Testament theme is picked up in the New Testament, though it takes a number of turns unanticipated in the Old Testament. The Gospel of John famously defines Jesus, the Logos, as comparable to the tabernacle: in the incarnation "the Word became flesh and made his dwelling among us" (Jn 1:14). Consequently, Jesus and the Gospel writers develop the idea of Jesus as the temple.[18] At the end of his ministry, in the upper room, he confides to the disciples that he is going away, and it sounds like the fall and the exile all over again—the presence of God being lost. Yet he quickly assures them that he will not leave them alone but will send the Comforter. At the same time he informs them that he is going to prepare a place for them so that they can all be together (note, not "I am going to redeem you or restore you so that we can all be together"). Even as he ascends to heaven several weeks later, he confirms that "surely I am with you always, to the very end of the age" (Mt 28:20). This is Immanuel Theology, with little reference to redemption or restoration.

---

[17]For development see John H. Walton, *Old Testament Theology for Christians: From Ancient Context to Enduring Belief* (Downers Grove, IL: IVP Academic, 2017).

[18]Nicholas Perrin, *Jesus the Temple* (Grand Rapids: Baker Academic, 2010).

Mere weeks later, another quantum leap in Immanuel Theology takes place as the Spirit descends at Pentecost. At the Tower of Babel the people wanted God to come down and take up his residence (rest) in the temple that they had prepared. That initiative was not focused on reestablishing the presence of God so that his name might be exalted, or reestablishing God's order, and it was therefore rejected. In contrast, at Pentecost God's presence descends and takes up residence (rests on; see Acts 2:3) in his temple (his people; 1 Cor 3:16; 6:19), and languages are united rather than divided. This is seen as a fulfillment of the promise (covenant; Acts 2:39) and results in all God's people returning home carrying God's presence with them rather than being dispersed with no presence of God as in Genesis 11. Certainly, Peter's sermon speaks of the salvation now made available, but it also indicates that the result of the forgiveness of sins is the presence of God (Peter says, "You will receive the gift of the Holy Spirit" [Acts 2:38], rather than addressing their redemption, destiny of heaven, or restoration of creation).

So it is today, and the mission of God to dwell among his people and be in relationship with them is realized in the church through the indwelling Holy Spirit. The mission of the church is similar to the mission of Israel. Like them, we are also being shaped as a community to host the presence of God by means of the Holy Spirit, who indwells us. We are redeemed indeed, by the blood of Christ, but our identity is not just as a redeemed community. We are the people of God who, as we co-identify with him, serve as his instrument, participating with him as he carries out his plans and purposes in the world. God's mission goes beyond redemption; it is focused on dwelling among the people he has created.

The anticipation of new creation stands as the climax of God's plans and purposes—his mission. The emphasis we find in Revelation 21 is not the redemption that we have received nor the restoration to Eden. New creation is the consummation of what was begun in Genesis 1–2 and interrupted in Genesis 3. God has been at work restoring his presence. The redemption he provided made renewed relationship possible and therefore was essential. The restoration of creation is also important because his presence is going to reach its full effect in a restored creation that is characterized by his perfect order. The description of new creation in Revelation 21 emphasizes God's presence: "And I heard a loud voice from the throne saying, 'Look!

God's dwelling place is now among the people, and he will dwell with them. They will be his people, and God himself will be with them and be their God'" (Rev 21:3). Non-order (such as the sea) and disorder have passed away. God's order is finally and fully established. The new creation has characteristics of Eden (Rev 22:2), and the curse has been resolved (Rev 22:3), but just as Jesus said of himself, "Something greater than the temple is here" (Mt 12:6), in new creation something greater than Eden is there. Relationship is fully open (Rev 22:4) and God's presence is fully accessible (no temple; Rev 21:22), and God's people have joined with him in his reign (Rev 22:5).

This survey of the theme of God's presence and his desire to be in relationship with his people demonstrates that the doctrine of creation itself, rightly understood, launches the mission of God. In response we need to give more attention to the theological trajectory of creation in the biblical view rather than being endlessly entangled in a supposed conflict between science and the Bible. The Bible is much more interested in creation as establishing God's presence than in the mechanisms used by the Creator. Science studies mechanisms; the Bible is more interested in agency—God as the Creator embarking on his mission to create people among whom he will dwell and who will be in relationship with him forever. Whatever mechanisms God may have used (and the Bible has little interest in them), creation should be the subject of our praise—a doxology directed toward the One who loved the very idea of humanity and created us with the intention of being with us. Immanuel Theology expresses how the mission of God focuses on presence and relationship: "I will dwell among you, and you will be my people, and I will be your God." This is why God created us. This is what the Bible is all about. This is what God has always wanted.

Part Three

The DOCTRINE of
CREATION PRACTICED

9

Intellectually Frustrated Atheists and Intellectually Frustrated Christians

The Strange Opportunity of the Late-Modern World

ANDY CROUCH

FOR A FEW YEARS I'VE BEEN EXPLORING a hunch that has two parts. First, the very best time to be an intellectually satisfied atheist was one hundred years ago, circa 1917, and it has been getting gradually harder ever since. And second, one hundred years from now, circa 2117, it is entirely possible that the most intellectually coherent account of the cosmos, on the basis of all that we will then know, will be in deep continuity with biblical, trinitarian faith.

Now, if that first claim is certainly contested and contestable, the second claim would seem laughably improbable to many of the most educated members of our culture in 2017. Nonetheless, I believe it could be true—but whether it actually turns out to be true depends a great deal on what choices we, as theologians and pastors, make right now, in 2017.[1]

---

[1] I would like to thank my colleagues at the John Templeton Foundation in 2017, and especially my wife, Catherine Crouch, professor of physics at Swarthmore College, for comments on early drafts of this paper.

THE TWELVE GREAT EXPLANANDA

Three fundamental realities require explanation, multiplied by four great mysteries that apply to each. Call these three multiplied by four the twelve *explananda* of human existence—the twelve questions to which any serious account of the world must offer some response, however tentative and humble.

The three fundamental realities are these:

1. The cosmos—the ordered and abundant reality that encompasses everything that can be perceived by human beings.

2. Life—any life, the mere fact of self-organizing, self-perpetuating systems of dynamic disequilibrium, of any kind or scale.

3. Humanity—human beings with our singular curiosity, capacity for knowledge, and understanding, authority, and vulnerability.

No one can claim to have given a satisfying account of the world as we know it without wrestling with these three distinct though profoundly interrelated realities.

And for each of these there are four great mysteries:

1. Their origin—why they came to exist at all.

2. Their sustained reality—the fact that not only did they begin to exist but they, at least for now, continue to exist.[2]

3. Their frustration—the incorrigible sense we have that they are not entirely what they were meant to be or, perhaps especially in the case of human beings, not *at all* what they were meant to be.

4. Their destiny—the end of the story, if there is any story at all, that will finally be told about each one.

I have called these four issues "mysteries" because they are not just "problems" with "solutions," or "questions" with "answers." They are, and have always been, far more than that. They are the concerns of the greatest human works of art, the various responses to the world that we call worship, including both praise and sacrifice, and our inarticulate sighs of wonder and groans of despair. To respond adequately to these mysteries is not to dispel

---

[2]These first two items have been classically treated together as the doctrine of creation, even though "creation" has come popularly to mean "origins" alone.

their mysterious quality; it is only to hold them together in some way that seems adequate to what we know about them, as well as what we do not know.

THE HIGH-MODERN EMERGING CONSENSUS

At the dawn of modernity, a novel and powerful way of responding to several of these twelve great explananda began to emerge.

It began to seem plausible that the cosmos itself had no origin—it was simply eternal, governed by a few immutable laws. In our corner of this cosmos, life had emerged, eventually including human life, powered and limited by the exact same laws. To be sure, how exactly to account for the origin of life was still a mystery, but two-thirds of the origin mysteries were solved: the cosmos itself, eternal and self-contained, needed no origin story, and human beings' origins simply were collapsed into the general story of the development of life.

The dominant *metaphor* for the world's origin and continued existence was *mechanism*. And the dominant *method* for understanding the world was *reduction*. The cosmos could be conceived as "nothing but" a mechanical set of atoms, points of mass bouncing off one another like billiard balls; or, in a slightly more sophisticated formulation, "nothing but" a set of fields interacting in elegant but entirely calculable ways.

Call this the "high modern" consensus about the world, the one that was really coming together as the twentieth century began. It was backed up by profound advances in our ability to formalize our knowledge of the world— above all, the tremendous predictive success of classical physics and the explanatory power of Darwin's evolutionary biology. It was also, from a certain point of view, a pleasingly closed system, entirely immanent, probably self-caused, and certainly self-perpetuating. And if there was still a great deal we didn't know, it seemed likely that the remaining gaps in this account would be filled in by scientific progress. We would progress from the deism of the eighteenth and nineteenth centuries, with its absent god, to a full-fledged atheism, with an unnecessary god. And as God dwindled away, the mysteries of the cosmos, life, and humanity would dwindle as well.

Richard Dawkins famously wrote in 1986 that Darwin "made it possible to be an intellectually fulfilled atheist."[3] Indeed, everything you needed to

---

[3]Richard Dawkins, *The Blind Watchmaker* (London: Penguin, 1986), 6.

be an intellectually fulfilled atheist was already present in 1917. Atheism even made geopolitical and historical sense, as Christian Europe slaughtered itself with mechanical efficiency in the trenches of the Great War. Not only was God not necessary as a hypothesis to explain the cosmos, but the absence of God seemed almost a necessary hypothesis to explain history and its frustration. The great explananda had collapsed into the machine. There were no mysteries after all, only the brutal necessity of history.

THE LATE-MODERN COMPLICATION

The ensuing hundred years have brought some rather extraordinary surprises. If I can give the label "high modernity" to the moment when all explananda seemed on the verge of being explained—indeed, explained away—I want to give the label "late modernity" to the way things have actually unfolded. In this era, the high-modern promise in certain key respects has unraveled, leaving us once again standing before some very powerful mysteries, this time not because we know too little about the world but because we know so much. The world turns out to be anything but a "nothing but" world.

There are four notable features of this emerging late-modern account of the world.

More contingent and dependent. First, in late modernity the cosmos turns out to be more contingent and dependent than we imagined in high modernity. It is almost certainly not eternal. This universe, we are sure beyond a reasonable doubt, had a beginning, 13.7 billion years ago. And because time is simply the fourth of the four dimensions of space-time, time itself had a beginning. Of course there are various proposals to embed our universe in either a sequence of universes or an abundance of alternative universes, but there are some powerful arguments suggesting that all these scenarios still require a beginning.[4] A universe with a beginning seems to need a cause—meaning we may need the hypothesis of a Creator outside of time, or some other fantastic hypothesis, after all.

But the cosmos we have also turns out to be contingent in complicated ways that we did not envision just one hundred years ago. In 1917 we could

---

[4]The arguments for a necessary beginning to the universe (and any conceivable alternative such as a multiverse) are described in Robert J. Spitzer, *New Proofs for the Existence of God: Contributions of Contemporary Physics and Philosophy* (Grand Rapids: Eerdmans, 2010).

conceive of the material world as made of a few elementary particles. But as the twentieth century went on, we would carry out increasingly ambitious experiments to find these elementary particles—and find out that there are not just a few of them but a rather bewildering profusion.

Furthermore, just in the past few decades we have realized that all of this matter probably amounts to only 7 percent of the matter and energy in the universe, with the rest—93 percent—being composed of dark matter and dark energy that does not interact with any of our experimental apparatus and for which we have at present no satisfying theoretical account. In the words of one physicist friend, "We have gone in my lifetime from thinking we knew about 80 percent of what there was to know about the matter in the universe, to thinking we know about 10 percent of about 7 percent of the universe."

And then it turns out that the equations that describe this model include a number of arbitrary constants that could have been otherwise—the G in the equation of gravity, for example. Tinker ever so slightly with many of these constants and we would simply not have a universe like ours at all. The vast majority of possible values for these constants yield a cosmos, if you can call it that, of "solitary primordial protons" that would drift through endless reaches of space, with two of them interacting once every billion years, and then only to bounce off each other into the endless emptiness.[5]

Indeed, the mathematician Roger Penrose has calculated the odds of the universe as we know it coming into existence to be 1 in $10^{10^{123}}$—a truly fantastic and inconceivably large number.[6] Just to count this number would require far more particles than there are in the universe. How did we end up, in the words of the astronomer Luke Barnes, with such a "fortunate universe"?[7] We are back in the realm of deep mystery. There is nothing mechanically certain about the cosmos we find ourselves in—quite the reverse. If the universe is so contingent and so dependent, on what is it dependent? Or on whom?

---

[5]"Galaxy Formation and the Fine-Tuning of the Universe for Intelligent Life," John Templeton Foundation, accessed December 2, 2017, www.templeton.org/grant/galaxy-formation-and-the -fine-tuning-of-the-universe-for-intelligent-life.
[6]Cited in Spitzer, *New Proofs*, 49.
[7]Geraint F. Lewis and Luke A. Barnes, *A Fortunate Universe: Life in a Finely Tuned Cosmos* (Cambridge: Cambridge University Press, 2016).

More relational. Second, the world is turning out to be far more relational than we would have thought in 1917. The metaphor of "particles" turns out to be mere metaphor. The metaphor of "fields" comes closer, but that does not quite do justice to the disorienting discovery that the things we think of as "things" in the world actually are deeply connected to and interdependent on one another. The phenomenon of quantum entanglement, in which two "particles" turn out to be able to communicate with one another instantaneously, suggests that the relationship between the two particles could be more fundamental than the particles themselves.

And we now can appreciate, in ways we could only dimly anticipate in 1917, just how connected is the world of living things. Duke theologian Norman Wirzba draws our attention to the writing of anthropologist Tim Ingold, who argues that rather than speaking of life as a *network*, in which discrete things enjoy some (perhaps optional) measure of connection to other things, it is more accurate to speak of life as a *meshwork*, a woven collection of relationships in which any given creature subsists. To really understand the way the living world functions, we have to move, Wirzba suggests, from "see[ing] the world as a collection of nouns [to seeing it as] a field of verbs."[8]

This is a reality for which mechanism and reduction are nearly useless categories, as Wirzba points out. "To understand life one must pay attention to the movement of interlacing. *Interlacing* is not the same thing as *interlocking.* The joining together that makes a knot is not simply a mechanical act. . . . The coming together of lines or strands that make a knot are much more intimate than that. To convey this intimacy, it is necessary to adopt the language of sympathy [or, in musical terms, polyphony]."[9]

More informational / logos-based. The world is far more contingent and dependent than we imagined; it is far more relational than we imagined; and third, it is far more informational than we imagined. The high-modern deist or atheist could, of course, see that the world had a kind of rational structure—that a few elegantly stated equations could give the terms of motion of both planets and charged particles. But I'm not sure that they could have anticipated how extraordinarily fruitful complex mathematics has turned out to be in describing the actual cosmos.

---

[8]Norman Wirzba, "Creation Through Christ," in *Christ and the Created Order: Perspectives from Theology, Philosophy, and Science*, ed. Andrew B. Torrance and Thomas H. McCall (Grand Rapids: Zondervan, 2018), 50.

[9]Wirzba, "Creation Through Christ," 49.

In 1917 a well-informed European might have heard of the mathematician David Hilbert and his concept of Hilbert space, a Euclidean space with infinite dimensions. They would also have been influenced by Hilbert's argument for formalism, the idea that mathematical truths need not refer to anything other than their own set of rules, certainly not to anything necessarily real in the natural or created world. Indeed, how could complex Hilbert space, which invokes imaginary numbers (based on the impossible square root of -1), be connected to the real world? And yet a few decades later, complex Hilbert space turned out to be essential for representing the state vectors required for quantum mechanics. Something that started out as pure mathematical exploration turns out to have incredible explanatory power for the world as we know it—a phenomenon Eugene Wigner called "the unreasonable effectiveness of mathematics."[10] The world, it seems, responds to and is built on some of the most elaborate ideas it is possible for human beings to rigorously formulate and entertain.

We also are beginning to understand how fundamental information is to the universe. Indeed, there is reason to think information is not merely a byproduct of material states but an independent feature of reality. The information theory pioneered by Claude Shannon has been essential to making progress on subjects like quantum entanglement. And information is encoded in life itself, in DNA, which turns out not to be merely a set of simple instructions like a rudimentary computer program for life but a kind of meshwork in itself that generates the astonishingly complex structures of folded proteins and that interacts epigenetically with its environment.

In 2017 we do not need to shy away from saying that the world has *logos*, elegantly structured information, at its heart.

More personal. A world that is more contingent and dependent, more relational, more profoundly rational and informational is ultimately a world that is more personal than we could have imagined. For persons are contingent, dependent, relational, logos-based, and logos-seeking creatures. And here we are. The world seems to have been in some way prepared for us. The fine-tuning that allowed galaxies to come into existence is far more exquisite

---

[10]E. P. Wigner, "The Unreasonable Effectiveness of Mathematics in the Natural Sciences: Richard Courant Lecture in Mathematical Sciences Delivered at New York University, May 11, 1959," *Communications on Pure and Applied Mathematics* 13 (1960): 1-14.

than that (so the philosopher Robin Collins has argued)—it actually is fine-tuned for the appearance of intelligent creatures that can actually apprehend and interpret the glorious simplicity and abundance of the world. The world is fine-tuned for the discovery of fine-tuning![11]

The evolutionary biologist Simon Conway Morris has argued that far from being a random walk through all possible outcomes, life itself converges inexorably on greater and greater structured complexity, and ultimately on rationality and self-awareness.[12] Evolutionary biologist Jeffrey Schloss traces the ways that life becomes more and more relational over time, more and more deeply capable of recognizing and bonding to another, and ultimately more and more capable of altruism, sacrifice, and love.[13]

Is it actually possible that an impersonal world could give rise to persons? Does it not make far more sense to suppose that such a world is in some mysterious way not just capable of personhood but defined by it, not just capable of life but defined by life?

SOMETHING-GREATERY

None of these features of our world was entirely unimaginable in 1917. But the dominant direction of science and intellectual life was reductionistic—or, to use the witty phrase attributed to G. K. Chesterton and others, "nothing-buttery." The findings of science in the last hundred years have done anything but confirm nothing-buttery. For those with eyes to see, they rather tend toward "something-greatery"—the sense that there is some greater reality, in some ways accessible to us and presumably in other ways entirely beyond our grasp, that has called such a world into being.

Even our contemporaries who are far from theism sense this. The entrepreneur Elon Musk suggested in 2016 that the chance that we are not living

---

[11]Robin Collins, "The Teleological Argument: An Exploration of the Fine-Tuning of the Universe," in *The Blackwell Companion to Natural Theology*, ed. William Lane Craig and J. P. Moreland (Chichester, England: Wiley-Blackwell, 2012).

[12]Simon Conway Morris, *Life's Solution: Inevitable Humans in a Lonely Universe* (Cambridge: Cambridge University Press, 2003).

[13]Jeffrey P. Schloss, "Would Venus Evolve on Mars? Bioenergetic Constraints, Allometric Trends, and the Evolution of Life-History Invariants," in *Fitness of the Cosmos for Life: Biochemistry and Fine-Tuning*, ed. John D. Barrow, et al. (Cambridge: Cambridge University Press, 2008).

in a computer simulation is "a billion to one."[14] This was actually subsequently ruled out by a rigorous mathematical analysis—it turns out no simulation could represent all the possible states of the universe.[15] But why is that any more plausible anyway than the belief that the contingent, relational, informational, and ultimately personal world is graciously created and sustained through a necessary, relational, personal Logos?

This does not mean that there will be no atheists in 2117 any more than that there were no Christians in 1917. In fact, just as most Westerners were still in some meaningful sense Christian in 1917, I fully would expect that most Westerners will still be in a most meaningful sense functionally agnostic, deist, or atheist in 2117. But the tide on Dover Beach will no longer be going out. It will be starting to rise.

And Yet . . .

So far this might seem like a reasonably conventional piece of Christian apologetics—if not wishful thinking.

But if Christian faith is going to be meaningful and persuasive in 2117, we need to take into account some other significant entries on the other side of the ledger. Because the truth is that the last century has also intensified the challenges for the would-be intellectually fulfilled Christian believer.

I want to group these challenges under two broad headings, which themselves are fundamental mysteries relating to our three great topics of human beings, life, and the cosmos. The two great mysteries are existence and death. Of course, these have been mysteries for all of human history. But they take on a new depth and urgency given what we know in late modernity.

Cosmic decay. Let's start with death or, more broadly, decay into disorder. The second law of thermodynamics had, of course, already been formulated by 1917. But in 2017 we see in clearer relief its universal applicability. Decay, in the sense of order giving way to disorder, is central to the functioning of the cosmos. And it has been going on from the very first moments of cosmic

---

[14]Rich McCormick, "Odds Are We're Living in a Simulation, Says Elon Musk," *Verge*, June 2, 2016, www.theverge.com/2016/6/2/11837874/elon-musk-says-odds-living-in-simulation.

[15]Zohar Ringel and Dmitry L. Kovrizhin, "Quantized Gravitational Responses, the Sign Problem, and Quantum Complexity," *Science Advances* 3, no. 9 (September 27, 2017), cited in Margi Murphy, "Scientists Confirm We Are NOT Living in a Computer Simulation," *Sun*, updated October 20, 2017, www.thesun.co.uk/tech/4594362/are-we-living-in-a-computer-simulation/.

history. Christians are comfortable, thanks to Romans 8, with speaking of the cosmos being held "in bondage to decay." But we have tended to treat this as the cosmic corollary of the death that came into the world through human sin. This cannot easily describe the steady increase in entropy that has characterized our universe—even as pockets of dynamic disequilibrium, up to and including life and human beings, emerged. The cosmos is being subjected to more and more futility each day, and it has been so from the beginning, not since some wrong turn caused by humans.

Death and predation. Likewise, we have overwhelming evidence that death and predation are more fundamental to life and its history than Christians have traditionally imagined. Creatures have been consuming other creatures' remains for two billion years and pursuing them and taking their lives for five hundred million. Indeed, without this extended cycle of life and death, we would not have the rich deposit of decomposition that we call topsoil, the few inches of earth on which human agriculture is utterly dependent. The relational nature of our existence, our dependent connection to countless other creatures, implicates us not just in the lives of other creatures but in their death and decay, which is the source of our life. This is not some recent aberration in the workings of our world—it is the living world as far back as we can see, and as far forward as we can peer. Is this level of decay, death, and predation really compatible with God declaring the world "very good"?

Violence. The last hundred years have complicated the human story as well. Human beings have known decay, predation, and death as long as we can remember, but the last hundred years brought violence on a scale never before seen in history. The scale of human destruction in the twentieth century utterly confounds the mind and heart. At least most of those killed in the Great War had been combatants subject to the laws of war. But already in 1917 the extermination of Armenians was under way. In the 1930s would begin the genocide of the Jews, God's chosen people. These were just two of dozens of systematic, vast efforts at eradicating human life that collectively took the lives of tens of millions of people. The twentieth century would go on to be more indiscriminately brutal than any other in history, and nothing has changed in the first seventeen years of the twenty-first.

What kind of god, we should seriously ask, would create a world that started to wind down the moment it was wound up, requires innumerable generations of creatures to give up their lives to others, and would tolerate the violation of his image on such a massive scale without seeming to speak a word or lift a finger in judgment?

Cosmic space and time. The deep mysteries of decay, death, and violence are paradoxically paired with another set of mysteries relating to existence. What there is seems far more permanently in bondage to decay than we might have supposed, and yet what there is also turns out to be far more extensive and abundant than we have ever imagined.

At the cosmic level, there is just a lot more stuff than we imagined. Six years after 1917, Edwin Hubble would observe a so-called nebula with enough resolution to discover that it was in fact a separate galaxy. A hundred years later, we believe galaxies number in the billions, perhaps one hundred billion—each one with billions of stars, unimaginably remote from our own sun. When God invited Abram to number the stars and promised that his descendants would be just as numerous, Abram might have imagined that there were thousands of stars. In fact, even if every human being who ever lived were to be a descendant of Abram, we would still be orders of magnitude short of that promise being fulfilled.

This cosmic abundance does not directly challenge biblical faith, to be sure. But there is something vertigo-inducing about discovering the vastness of the universe. The Bible is ultimately a human-scale story, with its climactic acts being God taking up residence among his people and then as a person. And yet we can now zoom out by power of ten after power of ten, and this story, so consequential to us, seems impossibly puny beside the universe. It is very hard not to feel, in Walker Percy's words, "lost in the cosmos."

Life on Earth and other planets. Then there is the matter of life. So far we live on the only planet we know of that is hospitable to life. But there are a few other planets out there. In 1992 the first exoplanet was discovered. As of October 19, 2017—and yes, this changes frequently enough that it needs to be dated—the total number of confirmed observed exoplanets was 3,532.[16] That is just the ones we have been able to detect. There are surely trillions of planets.

What are the chances that our planet has the only life in the whole universe? To believe that requires a kind of anthropic principle run in reverse.

---

[16]NASA Exoplanet Archive, https://exoplanetarchive.ipac.caltech.edu, accessed October 19, 2017.

It beggars belief. Who knows if we will ever encounter life in other solar systems—but the idea that it is not there seems incredible. And given that life seems to cascade upward in complexity given sufficient time and inputs of energy, eventuating in contingent, dependent, relational, logos-seeking, and logos-bearing creatures—why would that be any different in other places?

The abundance of history. One more mystery of existence relating to human beings is another mystery of abundance—in this case, the abundance of human history. This history is larger than the biblical imagination ever really seems to encompass. It is more extensive in space: during the whole redemptive history narrated in the Old Testament beginning in Genesis 12 and based in the Fertile Crescent, great civilizations were rising and falling in Asia and the Americas. And it goes back far deeper in time. Thirty thousand years ago, a community of artists painted at least thirteen different species of animals on the walls of a cave in modern-day Chauvet-Pont-d'Arc. These artists were unmistakably and unarguably human. Where does their history fit into our sacred history? This abundance of the human story confounds any too-small, too-literal rehearsing of the biblical story that seems almost entirely innocent of deep time.

A PASTORAL, INTERPRETIVE RESPONSE

This is our situation in 2017, poised between the satisfied atheism of 1917 and, possibly, the revived trinitarian belief of 2117. What we have is a bunch of intellectually dissatisfied atheists and a bunch of intellectually dissatisfied Christians. Or we would, if either side were intellectually honest.

In fact, if we were to pay full attention, we would admit that we are faced with very, very deep mysteries that our forebears in 1917, whether atheist or Christian, could not easily have imagined. I think that the outcome in 2117— whether my hopeful prediction about the renewed plausibility of biblical, trinitarian faith comes true or not—depends very much on what kind of pastoral and theological framework we bring to these mysteries. Especially our pastoral framework. There are substantive theological resources to address many of the issues I've raised. But they have not made pastoral contact with the great majority of the people of God.

So what's a pastor theologian to do?

Be orthodox. First, we should be orthodox. Nothing in the emerging scientific account of the cosmos, life, or humanity fundamentally challenges what Christians everywhere and at all times have believed. The high-modern capitulation in liberal theology, in which core orthodox doctrines were discarded in the face of the alleged transformative power of electric light bulbs,[17] was premature and a failure of nerve.

Be confident. Second, we should be confident. Not arrogant, not overconfident, not certain, but confident that the more we know about God's world, the more God's revelation of himself will be confirmed. We do not need to shrink back when children ask us questions about the origins of life or humanity or the cosmos. To the contrary, we should gladly explore with them the mysteries of the world.

Be humble. Third, we should be humble. Every human graced to know the truth about our universe can have no possible response but humility at the vast gift of cosmos, life, and humanity. How can we gaze at the astonishing images from the Hubble Extreme Deep Field project and still exalt ourselves?

Indeed, the way to truly be human—including a human who has access to all we know about the world—is to be humble. The people of God in the biblical narrative are not the emperors or the great civilizations that integrate technological prowess with extensive power. They are the ones who look up at the vastness of heaven and say, "What are human beings that you are mindful of them, / mortals that you care for them?" (Ps 8:4 NRSV).

Be revisionist—or "resurrectionist." But I do not think that being orthodox, confident, and humble is quite enough. We need one more quality, usually associated with an uncomfortable word: we need to be revisionist. I am fully aware how little evangelical Christians like that word. But it's just true. The late-modern account of the world requires substantial modifications to our tacit traditional worldview—including that held even by the greatest of previous generations of theologians. No one can really see their own worldview until it is cast into relief by another worldview, or by another view of the world. Now that we know the true extent of decay, death, predation, and violence, as well as the abundance of creation and history, we

---

[17]Rudolf Bultmann, *New Testament and Mythology and Other Basic Writings*, ed. Schubert M. Ogden (Minneapolis: Fortress, 1984), 4.

discover that we have unthinkingly held a view of the world, often with quasi-biblical sanction, that was far too small.

But there may be an alternative word that is better. Maybe what we are called to be is *resurrectionist*.

The resurrection of Jesus Christ, the firstborn of the new creation, should have already alerted us to the possibility that God's ways are far stranger and more wonderful than our ways. It set in motion a far-reaching cascade of re-examination of God's revelation that could ultimately lead the church to the astounding affirmation that Christ rose "in accordance with the Scriptures," even though no one looking at the Hebrew Bible, other than Jesus of Nazareth himself, had ever imagined the death and resurrection of the Messiah. The church emerged from its urgent searching of the Scriptures with a far more profound account of God's nature and purposes throughout history.

Among the most important implications was the rise of cosmic language for Christ's rule. In Isaiah, God had already said to his servant,

> It is too light a thing that you should be my servant
>> to raise up the tribes of Jacob
>> and to restore the survivors of Israel;
> I will give you as a light to the nations,
>> that my salvation may reach to the end of the earth. (Is 49:6 NRSV)

But the early church realized that it was too small a thing, in light of the resurrection, for God's salvation to reach to the end of the earth. Jesus is the firstborn of *all* creation, and he is the firstborn of the *new* creation. "He himself is before all things, and in him all things hold together" (Col 1:17 NRSV). The signature book of the last five hundred years for us Protestants has been Romans. The books for the next hundred years need to be Colossians and Ephesians.

RESURRECTIONIST THINKING IN ACTION

Let me give two brief examples of how this kind of revisionist, resurrectionist thinking might proceed.

Life. What if Christians became known for their bias toward the abundance of life in the cosmos? Not defensively exceptionalist, holding out some strange hope that our planet is the only fruitful planet of the trillions in the universe, but actively hopeful that because the Spirit is the Lord, the giver of life, and has hovered over all creation, we have every reason to expect that the vast expanse of galaxies are teeming with life? What if Christians urged their children to go into astrobiology precisely to advance this search as part of our image-bearing dominion over all other living creatures?

Let me be a bit more provocative. What damage, exactly, would be done if we discovered that God had created others in his image on other worlds? I actually think we have an analog for how to approach this in the cross cultural encounter that is the best form of Christian mission. The human family already has incredibly manifold expressions of the divine image in our different cultures. Now of course human beings share a common ancestry in a way that we would not with alien life. But why would we not hope and expect that, just as we find signs of the infusing of the divine image in every new cross cultural encounter, as well as its compromise through sin, we would find the same in encounter with other rational, ensouled beings? Why would Christians not urge their children to grow up to join SETI, the search for extraterrestrial intelligence?

Death. We also need a new perspective on death. Here again I am incredibly helped by the work of Norman Wirzba in his book *Food and Faith: A Theology of Eating*, in which he reflects carefully on the inevitable death and sacrifice that is involved in food.[18] This can be read in one way as alarming and tragic, but there is another way to see it: as a profound expression of love, in which creatures not only live their own natural lives but also give up their lives for the sake of the nourishing and thriving of others.

We lived for centuries in the church, especially at the level of lived lay theology, with a too-easy conflation of body (*soma*) and flesh (*sarx*). Could it be that just as the body is not merely flesh (*sarx*), so not all death is the sin-death that Paul talks about, the death that carries the sting of sin; that not all death is the "second death" of damnation and judgment that Revelation speaks of? Could it be that natural death and even much of what we call natural evil are somehow part of participation in the creative love of God?

We have an interesting clue here in the ambiguous grammar of Revelation 13:8, which used to be translated, "The Lamb slain from the foundation of

---

[18]Norman Wirzba, *Food and Faith: A Theology of Eating* (Cambridge: Cambridge University Press, 2011).

the world." Modern translations tend to refer that phrase "from the foundation of the world" to the Book of Life in which the names of the redeemed are written. But the phrase comes stubbornly at the end of the sentence in Greek, reminding us that the Lamb was slain by "the definite plan and foreknowledge of God," that in some real sense the Son has always been the self-emptying one who humbles himself even to death.

"Slain from the foundation of the world." What if all creation has in some mysterious way participated in the sacrificial self-giving of the Lamb? What if every exploding star, every falling leaf, every falling sparrow, every gazelle brought down by a lion, every child lost too soon, has not died in vain? What if their loss is in fact precious in the sight of God the Creator who does not forget, held even to the restoration of all things as a participation in the eternal self-giving of the Son? What if the Logos, the Word through whom all things are made, is fundamentally and essentially a sacrificial Word, a principle of endless giving love that does not fear death or decay, knowing that all things have been given into his hand, that he has come from God and is going to God? Then all things really do hold together—not just the easy or pleasant things, not just the baby star but the dying star—and the whole story of the cosmos from start to finish is just the first act in a story whose ultimate goal lies beyond the event horizon of the universe, whose meaning has nonetheless broken into this cosmos in the resurrection of the Son.

FAITH IN 2117

This is just the smallest glimpse of the kind of pastoral and theological rethinking that is ahead of us. Our doctrine of creation must not be reduced to a backward-looking tale of origins. It must be a forward-looking witness to the restoration of all things, rooted in God's raising Jesus Christ from the dead. If we do this work faithfully (and of course it is not just a matter of talk or speech but of deeds of power, love, and courageous sacrifice), the church of 2117 will have a doctrine of creation adequate to the cosmos and, even more important, will have a doxology of praise:

> You are worthy, our Lord and God,
> > to receive glory and honor and power,
> for you created all things,
> > and by your will they existed and were created. (Rev 4:11 NRSV)

Worthy is the Lamb who was slain,
to receive power and wealth and wisdom and might
and honor and glory and blessing! (Rev 5:12 ESV)

To the one seated on the throne and to the Lamb
be blessing and honor and glory and might
forever and ever! (Rev 5:13 NRSV)

Amen.

10

It All Begins in Genesis

Thinking Theologically About Medicine,
Technology, and the Christian Life

PAIGE COMSTOCK CUNNINGHAM

I N SEPTEMBER 2017, THE CASSINI SPACECRAFT ended its nearly twenty-year journey to Saturn, sending back some of the most spectacular photos of a planet ever seen. Artificial-intelligence systems can now perform tasks such as visual recognition, speech perception, and language translation. This powerful technology is as close as one's smartphone, or Amazon Echo. Advances in science and technology are mirrored by seemingly miraculous advances in medicine. Near-horizon medical technologies promise to restore sight to the blind and make the lame to walk.[1]

We live in a digitally mediated, technologically sophisticated, medically advanced, and scientifically awe-inspiring world. Yet not all these advances are beneficial to society. Robots threaten to replace thousands of jobs, including

---

I gratefully acknowledge my colleague Michael J. Sleasman, an essential and generous dialogue partner, for his significant influence on my theological arguments and his critical review of this essay. I am also grateful to Michael Cox, who provided critical review, helpful suggestions, and editorial assistance. All errors in judgment and conclusions are mine alone.

[1]Maya Wei-Haas, "Could This Bionic Vision System Help Restore Sight?," *Smithsonian*, October 19, 2017, www.smithsonianmag.com/innovation/could-bionic-vision-system-help-restore-sight-180965305; Tim Newman, "Stem Cell Research Offers New Hope for Restoring Sight," *Medical News Today*, January 11, 2017, www.medicalnewstoday.com/articles/315135.php; Sarah Kaplan, "These Monkeys Were Paralyzed Until a Brain Implant Helped Them Walk Again," *Washington Post*, November 9, 2016, www.washingtonpost.com/news/speaking-of-science/wp/2016/11/09/these-monkeys-were-paralyzed-until-a-brain-implant-helped-them-walk-again.

positions for whom there are no human applicants.[2] In Japan, robots are proposed as personal companions and nursing aids for the elderly.[3] Physicians and pharmacists are increasingly being pressed into service to address concerns that not too long ago would have been handled by others. Drugs are expected to assist in classroom management, to blunt undesirable personality traits such as extreme introversion, to avoid the effects of staying up all night, or to give a cognitive boost to a student with normal brain function.

More and more of life is being medicalized, with pharmaceuticals on hand to modify ever-expanding cognitive and behavioral variations.[4] Medical research aims not only at curing disease but also at discovering potentially lucrative aspects of postponing aging. Modern medicine offers to overcome human limitations through drugs, such as mood brighteners, brain boosters, alertness drugs, or memory aids. Most of these drugs have legitimate uses for diagnosed conditions. But when used by a healthy patient with no therapeutic need, or to compensate for behavioral choices, such as staying up all night playing video games, concerns naturally arise. These medical technologies blur traditional lines between *therapy* and *enhancement*, between the goals of medicine and the desires of consumers with the means to pay for medical technologies.

The traditional goals of medicine are undergoing a paradigm shift[5] away from the Hippocratic model, the physician-patient relationship based on trust, and the patient's confidence in the virtue and skill of the physician. The trend toward a provider-consumer relationship is increasingly at odds with Hippocratic, virtue-based medicine. Confidence in medical technologies threatens to eclipse the value of physicians' character, competence, and care.

MEDICINE AND THE TECHNOLOGICAL IMPERATIVE

Medical ethics, that is, clinical ethics and decisions at the bedside, is being transformed by a "second Copernican revolution," what Michael Sleasman

---

[2]Nigel M. de S. Cameron, *Will Robots Take Your Job?* (Hoboken, NJ: Wiley, 2017).

[3]Andrew Tarantola, "Robot Caregivers Are Saving the Elderly from Lives of Loneliness," *Engadget,* August 29, 2017, www.engadget.com/2017/08/29/robot-caregivers-are-saving-the-elderly-from-lives-of-loneliness/.

[4]Steven M. Meredith, Laura M. Juliano, John R. Hughes, and Roland R. Griffiths, "Caffeine Use Disorder: A Comprehensive Review and Research Agenda," *Journal of Caffeine Research* 3, no. 3 (2013): 114-30, doi:10.1089/jcr.2013.0016.

[5]Edmund Pellegrino, "Professional Ethics: Moral Decline or Paradigm Shift?," *Religion and Intellectual Life* 4 (Spring 1987): 21-22.

describes as the "technological turn."[6] Technology is interfacing with our understanding of being human in ways that could not have been anticipated. Medicine is adopting technologies that offer the ability not only to correct for normal functioning, or to repair an injury, but also to make us, in the words of Carl Elliott, "better than well."[7] But do they really make us "better" humans? How are we to think about these seemingly miraculous advances in medicine and technology? How then shall we live in the MedTech Age?

Sleasman goes on to observe that "while we may be finite embodied beings, the human imagination appears to have no limitations in its machinations to devise creative ways to dehumanize our existence in the proliferation of challenges facing us today."[8] Are we even aware of the dehumanizing potential of the technologies we employ? Sherry Turkle writes, "Technology is seductive when what it offers meets our human vulnerabilities."[9]

Observers of society, business, education, and psychology are among those who have begun to question our ready embrace of technologies, particularly digital communication and its pervasive presence in our lives, whether waking or sleeping. The smartphone has invaded the campus, the classroom, and our social life. Psychology professor Jean Twenge recently observed that the arrival of the smartphone has "radically changed every aspect of teenagers' lives, from the nature of their social interactions to their mental health."[10] The transformation is far from benign. The "iGen" generation is distinct in how they spend their time, in their views of religion, sexuality, and politics, and in their obsession with safety. They are also more anxious, lonely, and depressed.[11]

Near universal adoption of smartphones and other digital technologies illustrates the power of the technological imperative. This is the idea that technological progress is inevitable and unstoppable. Because a particular

---

[6]Michael J. Sleasman, "Bioethics in Transition: Framing the Discussion," *Dignitas* 21, no. 2 (Summer 2014): 5, https://cbhd.org/content/bioethics-in-transition-framing-the-discussion.

[7]Carl Elliott, *Better Than Well: American Medicine Meets the American Dream* (New York: Norton, 2004).

[8]Sleasman, "Bioethics in Transition," 1.

[9]Sherry Turkle, *Alone Together: Why We Expect More from Technology and Less from Each Other* (New York: Basic Books, 2011), 1.

[10]Jean Twenge, "Have Smartphones Destroyed a Generation?," *Atlantic*, September 2017, www.the atlantic.com/magazine/archive/2017/09/has-the-smartphone-destroyed-a-generation/534198/.

[11]Jean Twenge, *iGen: Why Today's Super-Connected Kids Are Growing Up Less Rebellious, More Tolerant, Less Happy—and Completely Unprepared for Adulthood—and What That Means for the Rest of Us* (New York: Atria Books, 2017).

technology enables us to do something—makes it possible—we *ought* to do it (the moral requirement), we *must* do it (the operational requirement), or we *will* do it (the inevitability requirement). In some ways technological progress is irresistible. The pursuit of determining the limits of possibility has led to making information storage smaller and smaller (remember the original floppy disk?) and making communication faster (remember using a dial-up modem?). Even now we become impatient when the dreaded "spinning wheel" on the screen signals a processing delay or worse. The sea changes in medicine and technology are of more than academic interest. They permeate the Christian community, raising serious concerns about how to use new technologies personally and for the benefit of others.

Pastoral Concerns in the MedTech Age

Perhaps these questions seem far afield from the concerns that cross a pastor's desk. But medicine and technology intersect with decisions that people make in their pursuit of health and other human goods. The underlying motivations, virtues, and theological and ethical considerations overlap. How many pastors would feel confident in responding to these scenarios?[12]

- Your son's teacher says she is noticing that he is distracted, can't sit still in his seat, and bothers other students. She wonders if he might have ADHD.

- An elderly man who has been through several rounds of chemo says he is ready to die and doesn't want any more treatments, but is afraid he might be sinning if he refuses chemo and feeding tubes.

- A friend in your women's Bible study group announces that her son and daughter-in-law are going to try IVF. They intend to donate any extra embryos to research.

- An acquaintance, who is a college professor, announces that he and his wife have decided their children will not have any vaccinations because of concerns about autism.

- A young couple comes to you. They've resolved to accept their inability to have children, but an older woman in their church has offered to be a gestational surrogate.

---

[12]With slight modifications, I have encountered each of these situations, usually by someone seeking advice.

These real-life situations are examples of bioethical dilemmas, ethical decisions regarding permissible and impermissible uses of the vast catalog of medical technologies and therapies. Bioethics deals not only with beginning-of-life and end-of-life concerns but also with issues arising throughout the lifespan at the intersections of medicine, science, and technology. Are pastors equipped to address these? Where do members of a congregation typically get their information and advice? Are congregants aware of ethical dimensions of these decisions?

Christians who consider using medical technologies, whether at the beginning of life or at the end, may be more greatly influenced by their culture than by their Christian worldview. While evangelicals hold a high view of Scripture, they may conclude from the lack of explicit biblical statements about modern technologies that "that which is not prohibited is permitted." The seeming absence of clear injunctions may cause some pastors to offer their advice as opinion rather than ethical guidance.

Ideally, Christians would consult their pastor before making a decision. Also ideally, pastors would feel equipped to provide moral guidance. In reality, pastors might not be consulted as often as they would wish, and they might be brought into a counseling situation after harmful decisions have been made.[13] I suggest that what is needed is *biblical wisdom* for ethical decisions about medically and technologically complex issues.

A Helpful Biblical and Theological Methodology

Is the Bible relevant? It goes without saying that modern technologies are not explicitly identified in Scripture. Tensions may arise when the Bible fails to provide explicit guidance, when it does not serve us as an answer book. Some Christians might conclude that biblical silence means that any technology or treatment that one can afford and that does not harm others is permissible, or at least that there is a great deal of ethical latitude. The lack of alignment between Scripture and today's technology, at first glance, does not mean that the Bible is irrelevant.[14] The inattentive reader might

---

[13]See the findings about pastors' confidence and desire to help in Paige Comstock Cunningham, "Becoming Bioethically Confident: The Contribution of Learning Experiences in Seminary and Congregational Ministry to Evangelical Pastors' Wise Leadership on Bioethical Concerns" (PhD diss., Trinity Evangelical Divinity School, 2017).

[14]See, e.g., Christopher J. H. Wright, *Old Testament Ethics for the People of God* (Downers Grove, IL: IVP Academic, 2004); Cyril S. Rodd, *Glimpses of a Strange Land: Studies in Old Testament Ethics* (Edinburgh: T&T Clark, 2001).

conclude that the Bible is ancient and foreign and thus of no value to us today. Yet if navigated carefully, the Bible is highly relevant to today's ethical questions.

Scripture can always illuminate, even when it does not provide a clear-cut answer. We are encouraged, over and over, to seek wisdom—biblical wisdom— to help us value all of life and to think and act theologically. God did not provide clear commands or instructions for every moral decision we might encounter. Rather than seeking the right stand to take or the correct clear-cut answer to the immediate dilemma, we are encouraged to become *morally reflective*, growing and maturing in our understanding and application of Scripture.

Dennis Hollinger, a Christian ethicist and president of Gordon-Conwell Theological Seminary, suggests that while the Bible may not be our *primary* source for doing ethics, it is clearly our *ultimate* source.[15] Its moral norms are the only completely reliable ones. The difficulty comes in the *interpretation* and *application* of those norms. When Hollinger first addressed the question of an evangelical approach to bioethics in the 1980s, he was responding to two dominant interpretive and methodological approaches: biblicism and ethical rigorism. Biblicism, sometimes called "divine command theory," is the tendency to "draw the content and style of ethical reflection directly from biblical statements, particularly imperative ones."[16] The tendency of a biblicist approach is to move straight from Scripture to a contemporary ethical issue, relying on imperative statements to the exclusion of others. Biblicism is also linked with literalism or proof-texting.

Ethical rigorism, "the application of moral principles in an absolutistic, noncompromising manner,"[17] taps into the yearning for simplicity and clear statements about right and wrong. It can even propose that the solution is "getting right with God" and then the ethical dilemmas will sort themselves out.[18] Its adherents tend to be selective in the issues they consider important and in the immoral behavior they identify, such as abortion, while ignoring others, such as poverty and war. The problem arises when there are genuine ethical conflicts, complexities, or ambiguities.

---

[15]Dennis Hollinger, "Can Bioethics Be Evangelical?," *Journal of Religious Ethics* 17, no. 2 (1989): 161-79.
[16]Hollinger, "Can Bioethics Be Evangelical?," 162.
[17]Hollinger, "Can Bioethics Be Evangelical?," 164.
[18]Wyndy Corbin Reuschling, *Reviving Evangelical Ethics: The Promises and Pitfalls of Classic Models of Morality* (Grand Rapids: Brazos, 2008).

Concerns about biblicism and ethical rigorism or absolutism remain. When Christians awaken to the moral implications of bioethical dilemmas, their initial tendency is to resort to these methodologies, seeking a Bible verse or two to affirm what they instinctively want to do, or asserting an absolute ethical principle to criticize someone else's decision. Perhaps a deeper problem is the outright neglect of theological thinking and ethical decision making, as well as a passive acceptance of cultural norms and expectations. A return to biblical and theological grounding is a restorative and practical corrective to enable better choices about life, death, and the moments in between. Theological doctrines, biblical principles, and biblical wisdom are relevant and essential for resolving contemporary bioethical dilemmas. Indeed, we are encouraged to seek wisdom.

> Tune your ears to wisdom,
>> and concentrate on understanding.
> Cry out for insight,
>> and ask for understanding. (Prov 2:2-3 NLT)

We can be confident that our search will not be fruitless. "If you need wisdom, ask our generous God, and he will give it to you" (Jas 1:5 NLT).

Theological doctrines and wisdom. In *Bioethics and the Christian Life*, David VanDrunen identifies a range of relevant theological doctrines to consider, such as God's sovereignty, providence, the goodness of creation, human agency, the *imago Dei*, and the contingent, finite, and limited nature of humanity.[19] Included among the more familiar doctrines is the long-neglected doctrine of suffering, which is a feature of some of the more difficult bioethical dilemmas. Suffering is inevitable and, at times, incomprehensible. Like Job, we may not understand the reasons why we or loved ones suffer. God may heal through modern medicine, by a miracle (and modern technology seems to be closing the gap between the two), or not at all. In the midst of suffering that seems irrational, God's grace is sufficient. God's purpose may be to refine faith, to conform us to the image of Christ, or simply to comfort us in our suffering.

A motivation of virtue and a heart cultivated in wisdom are dispositional necessities in the context of resolving a bioethical dilemma. Although much

---

[19]David VanDrunen, *Bioethics and the Christian Life: A Guide to Making Difficult Choices* (Wheaton, IL: Crossway, 2009).

of bioethics is about actions, virtue enables the actor to be the kind of person who will carry out the right actions. It is "a character trait that orients or disposes a person to act in a good way."[20] The pursuit of wisdom arises from the desire to understand reality, how the world works. It means to cultivate an ethics that is, as John Kilner puts it, "God-centered, reality-bounded," one that takes into account the present reality of the created world, the God who made it, and the people who live in it.[21]

Wisdom gives guidance in understanding and applying Scripture. Hermeneutical resources are available for the reader who desires to be a "wise and faithful interpreter" and to "develop the habits and practices required to be good readers of Scripture."[22] It helps us understand how to attain moral goals. It requires exercise of discernment, a task that may be uncomfortable for those who prefer black-and-white ethical solutions. Yet Scripture is silent on medical technologies such as in vitro fertilization, chemotherapy, Ritalin, genetic engineering, or feeding tubes. Biblical wisdom is foundational to wise, good, and right decisions. Biblical wisdom is found not only in the Wisdom literature of Job, Psalms, Proverbs, and Ecclesiastes but throughout the Scriptures, beginning with the core narratives of Genesis.

Biblical wisdom is a richer way of answering life questions than biblicism or ethical absolutism. It grapples with the way things are, with reality. It takes seriously the truth that God created the world and that he is sovereign over all creation. Wisdom helps us understand both our physical and social world. The root of wisdom is reverence and commitment to God. It means that we have a responsibility to seek out God's truth about how the world is structured and to live within the harmony established by God, in pursuit of shalom. Human resistance to these realities and truths and the consequent rupture of shalom are seen clearly in biblical narratives, beginning with the stories in Genesis.

Genesis narratives are down-to-earth, unflinchingly depicting flawed characters: deceivers, cheaters, adulterers, murderers, and scoundrels. The

---

[20]VanDrunen, *Bioethics and the Christian Life*, 69.

[21]John Kilner, *Life on the Line: Ethics, Aging, Ending Patients' Lives, and Allocating Vital Resources* (Grand Rapids: Eerdmans, 1992), 22.

[22]"About *Hermeneutics as Apprenticeship*," Baker Publishing Group, www.bakerpublishinggroup .com/books/hermeneutics-as-apprenticeship/347610, accessed December 4, 2017. David I. Starling, *Hermeneutics as Apprenticeship: How the Bible Shapes Our Interpretive Habits and Practices* (Grand Rapids: Baker Academic, 2016).

story lines are not simple but are filled with complexities that portray the way things were in these characters' lives. The wisdom in Genesis and throughout Scripture leads us to acknowledge the reality of how things are and to live life well.[23] It is wisdom that stands in contrast to the superficiality of some of our pursuits.

Contemporary technologically enhanced pursuits tend to be superficial: the superficial beauty of Botox and facelifts, the superficial wellness of yoga and Paleo, and the superficial relationships of Facebook and Instagram. They are superficial because they are, sometimes literally, skin deep. We can use social media to enhance our reputation, carefully choosing what we will post. These pursuits are tempting, in large part because of the truth that underlies each of them. We are created for beauty, for human flourishing, and for relationships that give purpose and meaning. Exhaustive and exhausting efforts to control one's appearance and reputation obscure clear vision about how these intermediate goals fit into God's design for human flourishing.

Consumption of technology in pursuit of a well-curated life can lead to counting on technology to alleviate all pain and suffering. Rather than the unrealistic pursuit of lives that are pain free, comfortable, and under our control, might we consider instead a well-lived life? Wisdom counsels that there are things in life that we need to grapple with and we need to accept by faith. The grappling and acceptance by faith of unresolvable problems are best framed by an attitude of gratitude and contentment.

Wisdom helps the Christ-follower to put the picture together in a new way, to see God *in* and *through* the complexities of life. It helps people deal with the pressure to be perfect and in control, a pressure that technology intensifies. Wisdom forces us to ask discomfiting questions. These are questions that do not readily yield an answer, questions such as asking God what he wants *for* us, and what he wants to *show* us.

Reading Genesis. Genesis can be read through many lenses. In light of the perplexing questions of medicine, technology, and health, in what follows I will attend to the Genesis account of creation, its theological anthropology (who we are as human beings), and the various narratives of relationality—between God and humankind, among human beings, and between human

---

[23]Dennis Bratcher, "The Character of Wisdom: An Introduction to Old Testament Wisdom Literature," Christian Resource Institute, March 2013, www.crivoice.org/wisdom.html. Accessed April 27, 2018.

beings and the created order. These themes resonate with contemporary challenges in medicine, technology, and the Christian life.

The creation account in Genesis is unique in ancient literature. In stark
contrast to the cosmic battle narratives of ancient Near Eastern mythologies,
creation in Genesis is peaceful. God speaks all things into being. God brings
order out of chaos. The creation narrative progresses from the broad to the
specific, establishing a trajectory of development and growth. The God who
creates is transcendent, separate from all creation. The statement seems almost
pedestrian in its familiarity, and yet it has staggering implications for wisdom
about medicine and technology. Medical technologies both tempt and enable
us to claim godlike powers, rejecting God's transcendent power and sovereignty, whether in the attempt to create "better" children through permanently
changing human DNA at the embryonic stage, or to avoid loss of control at
the end of life by demanding that doctors assist in planned deaths. But in
creation we see a God who pronounces that the world is good and that humankind is very good, not in need of revision. We see a God who is sovereign
over life and death, who provides for human needs even when his people rebel
and reject him. Most importantly, we see a God who gives because he loves,
and who cares for us with the watchful tenderness of a good shepherd.

Relationship between God and humankind. Humankind occupies a distinct place as the pinnacle of creation. Alone among all species, humans are
created in the image of God (Gen 1:26-27). Given the scant amount of attention in the text, it is not surprising that the meaning of "image of God"
has been debated throughout the history of the church.[24] Whatever else it
means, "image of God" makes it clear that humankind is distinguishable
from the rest of creation. We also are characterized by creatureliness.
Humans, despite everything more that we are, are fully a part of this created
reality. We are finite, existing in time and space, limited both physically and
morally, and contingent in our dependence on God for our existence and
ongoing life. We did not cause ourselves to exist. Our limitations mean that
there are physical limits we cannot exceed and that it is impossible for us to
know everything. At the very least, this counsels epistemic humility that
generates an ethic of restraint.

---

[24]See, e.g., the recent discussion in Richard Lints, *Identity and Idolatry: The Image of God and Its
Inversion*, New Studies in Biblical Theology (Downers Grove, IL: IVP Academic, 2015).

Relationships among human beings—Sarah, Hagar, and Abraham. The Genesis narrative moves from creation by God to the lived reality of the people he made. Sin marred the goodness of the natural world and of human beings. Iain Provan suggests that the curse that fell on Eve of greatly increased pains in childbearing (Gen 3:16) may be understood to encompass also the pains of family life in general.[25] Might this not also include the suffering of childlessness?

The curse of the barren womb is introduced in the story of Sarah and Hagar. The account maps the difficult, problematic, and enduring consequences of pursuing the blessing of a child by taking matters into one's own hands and introducing a third party into the marital relationship. God's promise to Abraham to make his descendants a great nation (Gen 12:3) languished for ten years, well past Sarah's childbearing years. In desperation Sarah may have reasoned that God's promise to make Abraham's descendants a great nation, through a child of his own body (Gen 15:4), had not specifically included a child of *her* body. She devised a plan to obtain a child, offering her maid as the means through whom she and Abraham might obtain children (Gen 16:1-2). Sarah's offer would have been a culturally expected choice.[26] The social consequences of this choice, however, were grave, for parents, children, and their descendants. Hagar's fertility, contempt, and subsequent insolence after Ishmael's birth aroused Sarah's jealousy and anger. Hagar fled prior to Ishmael's birth and, after Isaac's birth, was cast out with her young son. Two sons sharing the same father and different mothers could not be raised in the same household. Might not Hagar's story parallel that of contemporary surrogates? Their body is employed to benefit another, they are vulnerable to exploitation and control, and their value diminishes after they have given birth. The tensions and conflict within the household over infertility and childbearing reappear in the story of Jacob's two wives and their servants.

---

[25]"Family life will be tougher because life in general will be tougher." Iain Provan, *Seriously Dangerous Religion: What the Old Testament Really Says and Why It Matters* (Waco, TX: Baylor University Press, 2014), 117-18.

[26]Neff observes that "according to Hurrian custom, when a marriage does not produce children, the wife is required to find a substitute." Hagar's child would have been Abraham's legal heir. Robert W. Neff, "Annunciation in the Birth Narrative of Ishmael," *Biblical Research* 17 (1972): 53.

The tensions of these ancient stories resonate with the stories of many in our day: infertility, the desire to have "a child of one's own,"[27] control over reproduction, and failure to trust God in seemingly hopeless situations. Today, a complex array of assisted reproductive technologies (commonly referred to as "ART") enables couples and individuals to reproduce, without demanding a respectful examination of the consequences or even awareness that these technologies are fraught with moral concerns. Introduction of third parties, whether through egg donation, sperm donation, or gestational surrogacy, is all too often viewed as a matter of physical, technological, or financial limits, rather than one of moral harm. Taking matters into our hands may reveal broken relationships with God and may harm our relationships with others, not the least of whom is the child.

Relationships between human beings and the natural world. Technology's influence, however, is pervasive and relentless in its demands. Another ancient story from Genesis connects human nature, sin, and the seduction of the misuse of God-given creative powers and knowledge. The Tower of Babel illustrates a cycle that is repeated over and over. God blesses people with his presence and gives them every good thing. Eventually, people fail to trust God—their Creator—and to enjoy the good he provides.[28] Just prior to this story is the flood account, which culminates in the rainbow and God confirming his covenant with Noah and all people: "Then God blessed Noah and his sons, saying to them, 'Be fruitful and increase in number and fill the earth'" (Gen 9:1).

After the flood, humanity did not fill the earth. Instead, they congregated together, built a city, and embarked on a massive construction scheme, the Tower of Babel. At least four overlapping problems are evident: pride, disobedience, abuse of knowledge, and the pursuit of control. The hubris of the tower builders foreshadows our own confidence in medicine and biotechnology, broadly construed. The ziggurat was a vivid enactment of their desire to control their own destiny, their effort to protect their place and reputation, and their impulse to protect themselves from dispersal and diminishment. They refused to multiply and fill the earth. Their unrestrained imagination and

---

[27]Gilbert Meilaender, "A Child of One's Own: At What Price?," in *The Reproduction Revolution: A Christian Appraisal of Sexuality, Reproductive Technologies, and the Family,* ed. John F. Kilner, Paige C. Cunningham, and W. David Hager (Grand Rapids: Eerdmans, 2000), 36-45.

[28]Allen Verhey, "The Cultural Geography of Cloning," in *Cloning: Christian Reflection* (Waco, TX: The Center for Christian Ethics at Baylor University, 2005), 11-21.

creativity, given as good powers, developed in a distorted direction. The abuse of technological creativity can unleash horrors on human beings and our environment. Through God's gift of knowledge, these people were able to make bricks, yet misuse of technological prowess led to their downfall.[29]

Knowledge today is increasing exponentially. Remarkable discoveries in medicine, science, and technology are such a regular occurrence that they seem commonplace. Many of these are directed toward good ends, but much of the progress is diverted toward ends that will maximize commercial potential, enhance personal choice, exert eugenic control, and benefit the already privileged. The pride and rebellion of the tower builders is echoed in modern technological attempts to overcome natural human limitations, for example, our need for sleep or the degeneration of aging bodies. Some technologies facilitate one generation exercising immense power over future generations, for example, through genetic engineering of DNA, which is irreversible. Or they fail to fulfill their promise of greater health, freedom, or comfort, trapping the consumer of the technology into an endless, unsatiated desire for more. As C. S. Lewis said, both science and magic are attempting to solve the same problem: "how to subdue reality to the wishes of men."[30] We cannot use medicine and technology to erase the reality of our finite, contingent, and limited creation.

The revolution in medical technologies, and all the power it has given us, raises concerns about what is desirable for human flourishing. Gerald McKenny retrieves questions first raised by Plato about the role of medicine and the pursuit of health: "What limits should we observe in our efforts to remove causes of suffering and improve bodily performance?"[31] Randall Soulen expands on McKenny's question and asks, "What is the place of health and illness in a morally worthy life? Above all, what is the telos, aim, or goal of life in light of which we can say what ought to be valued about human life, and what the role of health, body, and medicine in it might be?"[32]

---

[29]See, for example, Graham Houston, *Virtual Morality: Christian Ethics in the Computer Age* (Leicester, England: Apollos, 1992), 69; C. Ben Mitchell and D. Joy Riley, *Christian Bioethics: A Guide for Pastors, Health Care Professionals, and Families* (Nashville: B&H, 2014), 195-96.

[30]C. S. Lewis, *The Abolition of Man* (New York: Macmillan, 1947), 88.

[31]Gerald McKenny, *To Relieve the Human Condition: Bioethics, Technology, and the Body* (Albany: State University of New York Press, 1997), 1.

[32]R. Kendall Soulen, "Cruising Toward Bethlehem: Human Dignity and the New Eugenics," in *God and Human Dignity*, ed. R. Kendall Soulen and Linda Woodhead (Grand Rapids: Eerdmans, 2006), 104.

The original goals of medicine centered on the healing of disease and al-leviation of suffering when possible and care for the patient always. Today the boundaries of medicine have expanded to encompass a broad notion of health or "bodily excellence."[33] More and more of life is being medicalized, from the classroom behavior of young children to the narrowing of desirable personality traits to enhancement of physical and cognitive performance. Ever since the Baconian revolution, our attempts to master nature have es-calated, limited only by the demands that autonomy be protected, that harms and risks be fairly predicted, and that the new technology be justly distributed.[34] This focus on mastery is inadequate for evaluating claims about the purpose or goal of human life, or about appropriate guidelines for and limits on medicine and biotechnology.

With a flexible and continually eroding notion of human dignity, there are few commonly accepted boundaries limiting what technology and med-icine can do to, and for, human beings. Kendall Soulen and Linda Woodhead contend that "the concept of human dignity is surprisingly fragile in this respect: it can be robustly maintained only within the context of a vision of reality that revolves centrally around something other than and greater than the dignity of the human being."[35]

Relationships between humankind and the natural world. The Genesis account establishes the foundation for creation care. We have theological reasons (creational finitude), ethical reasons (sufficiency), and virtuous reasons (self-restraint and frugality) to care for the earth.[36] William Dyrness argues that the natural characteristic of the earth is goodness and fertility, yet rebellion against God disrupts people's relationships not only with God but also with the land. Rather than goodness and fertility, the land becomes cursed and barren.[37]

Steven Bouma-Prediger argues that Christians' responsibility for cre-ation care begins with knowing where we are, becoming literate about our

---

[33]Soulen, "Cruising Toward Bethlehem," 105.

[34]McKenny, *To Relieve the Human Condition*, 8.

[35]R. Kendall Soulen and Linda Woodhead, *God and Human Dignity* (Grand Rapids: Eerdmans, 2006), 14.

[36]Steven Bouma-Prediger, "Creation Care and Character: The Nature and Necessity of the Ecological Virtues," *Perspective on Science and Christian Faith* 50, no. 1 (1998): 6-21.

[37]William Dyrness, "Stewardship of the Earth in the Old Testament," in *Tending the Garden*, ed. Wesley Granberg-Michaelson (Grand Rapids: Eerdmans, 1987).

environment. We live on the earth, a planet with magnificent, spectacular, harsh, strange, terrible, and beautiful landscapes. Even with general knowledge of the earth, we may not recognize the particular migratory birds, spring flowers, or nighttime mammals in our neighborhood. This kind of particular knowledge is significant, because "we care only for what we love. We love only what we know. We truly know only what we experience."[38] Genesis 6–9 reminds us that God made an unchanging covenant with the earth and its creatures, never again to destroy them by a flood. God, not humankind, is at the center of all things (Job 38–41). In Christ, all creation is held together, and through him creation will be redeemed by its renewal (Col 1:15–20). Bouma-Prediger reasons that the future God intends for us is good. Revelation 21–22 portray a world of shalom, "an earthly vision of life made good and right and whole. Heaven and earth are renewed and are one."[39]

Charles McCoy suggests that the world was created good and is still unfolding and being renewed as we move toward completeness and the eschaton, a future that is oriented toward, and rests in the hands of, God.[40] He emphasizes the symbiotic nature of creation, stating that acts that violate the covenant do harm to every part of creation. Thus, we have a responsibility to ask questions about our use of biomedicine and biotechnology that extend beyond their immediate effects on our own body. For example, synthetic hormones, improperly disposed prescription drugs, and vitamins that are eliminated from the body accumulate in harmful levels in the public water supply. Individual autonomy must be exercised in light of our common humanity and the cumulative effect of individual decisions.

The Genesis account of Noah and the flood introduces the use of animals for human well-being, when God gave Noah "everything that lives and moves" as food, but with the caution that an accounting would be demanded for their lifeblood (Gen 9:1-7). Whether for food or research, we must not callously use animals. For example, medical research ought to be conducted with great

---

[38]Steven Bouma-Prediger, *For the Beauty of the Earth: A Christian Vision for Creation Care*, 2nd ed. (Grand Rapids: Baker Academic, 2010), 21.

[39]Bouma-Prediger, *For the Beauty of the Earth*, 109.

[40]Charles S. McCoy, "Creation and Covenant: A Comprehensive Vision for Environmental Ethics," in *Covenant for a New Creation*, ed. Carol S. Robb and Carl J. Casebolt (Maryknoll, NY: Orbis Books, 1991), 212-25.

care for animal welfare, avoiding the infliction of unnecessary pain. We respect both the goodness of the animal and our moral responsibility as their stewards. Our stewardship points us back to creation, reminding us that we are entrusted with God's care for the creatures and creation he loves.

Toward Shalom: Doxological Gratitude and Contentment

The antidote, if you will, for the desperation that Sarah faced, or the pride of the tower builders, is a different understanding of human flourishing. Yet today we face the same temptations they did to use relationships or abuse creative gifts to get what we want—to use technology to transcend our human creatureliness.

We are the most materially privileged culture in history. Why then do we seem to be so unhappy? Each new technology is accompanied by the promise of more time and simplification. Yet people are busier than ever and in search of places of retreat. In *Beyond Therapy: Biotechnology and the Pursuit of Happiness*, the President's Council on Bioethics wrestled with the reality that technology drives toward more and more improvement of human beings, and with the unanswered question, *What is it about human beings that needs improvement?*[41] Psychiatrist Monty Barker writes, "What is the difference between striving for excellence and seeking perfection? The former is attainable, the latter is not. The former spurs us on, the latter so often leads to chronic frustration, even despair and depression."[42] This is *not* a path toward human flourishing.

Excessive attention to health and human flourishing risks the idolatry of making health the highest good. A person can be sick, even dying, and still flourish. Being reminded of one's mortality, as Allen Verhey comments, helps one to see things more clearly. He observes that human flourishing "depends first of all and fundamentally upon God . . . and upon the ways in which this God relates to us as the One who creates, as the One who draws us toward God's good future, and as the One who reconciles us when we are estranged from God, from [our] neighbors, and from the creation."[43]

---

[41]Leon Kass, *Beyond Therapy: Biotechnology and the Pursuit of Happiness* (New York: Harper Collins, 2003).

[42]Monty Barker, endorsement for Richard Winter, *Perfecting Ourselves to Death: The Pursuit of Excellence and the Perils of Perfectionism* (Downers Grove, IL: InterVarsity Press, 2005), 1.

[43]Allen Verhey, "Theological Dimensions of Health and Human Flourishing in a Fallen World"

God creates, sustains, and loves us. "With hospitable generosity God gives creatures the time and space to be themselves," says Verhey. God freely loves, freely gives. We are free to flourish. Our proper response is one of doxological gratitude. Verhey steers gratitude away from present suffering and toward eschatological hope. God did not simply create in the past. He is also drawing all things toward God's good future. "Our bodies have a telos that surpasses our imagination, but that nothing can destroy or wither." In a gift of divine freedom and divine love, God sent his Son, expressing "the eschatological glory of the Father." Jesus established a new community, one that we continue as we feed the hungry, share resources with the poor, and care for the sick.

In persistent generosity, God heals our separation and reconciles us in our estrangement from him, from each other, and from his created world. Apart from his self-giving love, apart from his generous forgiveness and reconciliation, we could not flourish. But because he freely gives, we are able to freely receive, in a desire for communion with God and with fellow human beings. Verhey calls our response "affective affirmation." In affective affirmation, we look on the other with joy, seek the well-being of our neighbor, and rejoice in the beloved. Verhey concludes, "In doxological gratitude, joyous hopefulness, and affective affirmation, we can flourish in sickness and in health, when we need care and when our care is needed."

How do we make sense of the cornucopia of technological options on display—from the amazing photographs taken by the *Cassini* spacecraft to my iPhone's ability to converse, take orders, and provide directions? How do we judiciously employ an array of medical technologies—from Botox to retinal implants, from Ritalin to manipulating the DNA of embryos? And how do we do so in the service of human flourishing? These serious, sobering questions demand thoughtful evaluation of their costs, risks, consequences, and proposed benefits. They require of the church something more than a culturally driven response. How *now* shall we live?

For pastors this is not simply a unique opportunity. Pastors have a responsibility to cultivate biblical wisdom in themselves and their congregations,

---

(plenary address, Center for Bioethics & Human Dignity, Deerfield, IL, July 19, 2013). An excerpted version is available at https://cbhd.org/content/cult-human-health. This was his last public address before his death in 2014, after a long struggle with amyloidosis. Quotations in the next two paragraphs are from this address.

oriented toward thinking and living theologically in the highest sense. A refined hermeneutic can enable the church to become wise and faithful interpreters of ancient texts, integrating orthodoxy and orthopraxy. These ancient
texts highlight the reality that our forebears in the faith faced. But our own
path of flourishing takes us through new challenges made possible by technological and medical advances. Following Jesus on this path is not for the
faint of heart, but it is a noble and rewarding journey. In order to walk this
path faithfully, we must seek biblical wisdom with epistemic humility—in
submission to the guidance of the Holy Spirit. In this way, the ancient and
foreign text can be a light to our contemporary path. May the Lord grant us
wisdom as we join his work of making all things new.

11

Justice, Creation, and New Creation

In Christ All Things Hold Together

KRISTEN DEEDE JOHNSON

A LITTLE OVER A DECADE AGO, I began to notice just how many evangelical Christians were talking about justice. Perhaps because I'd just returned from four years in Scotland, this struck me as a dramatic shift within a short period of time. Truth be told, I am almost innately nervous about popular and trendy things. Whether it's the latest style or the newest theological trend, my response is caution. As someone whose early academic formation was in sociology, I want to understand why—Why this shift? Why now? What larger cultural reality is this reflecting? And as someone whose later formation was in theology, I want to ask how this fits with Scripture and the long convictions of the Christian tradition.

My sense was that this seemingly newfound passion for justice was actually ancient. My hope was that what we were seeing was a recovery of something intrinsic to the Christian faith that, for particular historical reasons, had been lost to some strands of Christianity in recent decades. This also raised a concern. If this was indeed the recovery of something that had been lost for a season, that might mean that contemporary resources that could help root this new passion for justice in Christ would be scarce.

As I surveyed the landscape, I found myself sharing some of the worries of Christian ethicist Stanley Hauerwas. Writing in the early 1990s to a more progressive segment of the American church that was marked by its own passion for justice, Hauerwas observes, "If there is anything Christians agree about today it is that [their] faith is one that does justice." But, he continues, in "the interest of working for justice, Christians allow their imaginations to be captured by concepts of justice determined by the presupposition of liberal societies." The problem: Christians "forget that the first thing as Christians [they] have to hold before any society is not justice but God."[1]

With these words, Hauerwas is reminding us that if we as Christians are to care about justice, it should be because that commitment to justice flows from God and is shaped by God. In other words, our imaginations for justice ought to be captured more by what God has revealed to us in Christ and Scripture than by cultural notions of justice.

This is not because we have nothing to learn from others who care about justice but because notions of justice drawn from other traditions are not as compatible as we might think. Here Hauerwas has been influenced by the thought of Alasdair MacIntyre, who notes that in Western society modern and individualistic conceptions of justice are placed alongside traditional, Aristotelian, and Christian conceptions of justice, even though these conceptions are not consistent with one another. The result, MacIntyre argues, is that "we have all too many disparate and rival moral concepts, in this case rival and disparate concepts of justice."[2] Like Hauerwas, I find MacIntyre's arguments both persuasive and significant. As Christians, we ought to be able to articulate the source of our conceptions of justice in Christ and Scripture as we seek to live according to God's call to justice.

As I listened to Christians describe their zeal for justice, it was, if I'm honest, sometimes hard to tell what conceptions of justice were shaping them. I do believe that these Christians had a sincere desire to connect their passion for justice with their faith in Christ, but they weren't able to articulate that connection. (I want to make sure to say that I view this as a result of the church's struggle to provide adequate biblical and theological formation rather than as the fault of these individual Christians.)

---

[1]Stanley Hauerwas, *After Christendom: How the Church Is to Behave If Freedom, Justice, and a Christian Nation Are Bad Ideas* (Nashville: Abingdon, 1991), 45, 68.

[2]Alasdair MacIntyre, *After Virtue: A Study in Moral Theory*, 2nd ed. (Notre Dame, IN: University of Notre Dame Press, 1984), 252. See also Alasdair MacIntyre, *Whose Justice? Which Rationality?* (Notre Dame: University of Notre Dame Press, 1988).

When talking about justice, many Christians I spoke to referenced prophets like Isaiah or Micah. Or they would cite the life and example of Jesus, noting how Jesus spent his ministry with sinners and tax collectors, spoke scathing words to those who were privileged, and began his earthly ministry by proclaiming, in the words of Isaiah, "The Spirit of the Lord is upon me, because he has anointed me to bring good news to the poor" (Lk 4:18; Is 61:1 NRSV).

The life of Jesus, as the Word made flesh, does have a lot to teach us about what matters to God. As people called by God to speak the words of God, the prophets likewise can show us what matters to God. But the words of the prophets and the life of Jesus represent only a portion of the Bible. What would we learn about God's justice and the calling God places on us, his people, to seek justice if we immersed ourselves in the whole story of Scripture? That was the question that ultimately guided my coauthor and me as we wrote *The Justice Calling*.[3] Sensing that the biggest need was for a *why* book (why is it that we're called to seek justice?) rather than a *how* book (how do we go about pursuing justice?), we decided to walk ourselves and our readers from Genesis through Revelation to explore what God has to say about why we should care about justice in this world.

But discussing the entire Bible presented a theological quandary. My training as a theologian shaped me to believe that we ought to think christologically about everything, even the parts of the biblical story that seem to precede Jesus Christ chronologically. My years studying the thought of Augustine of Hippo further convinced me that Jesus Christ is essential for real and true justice.[4] So as we told the biblical story as it unfolds in the canon, I tried to connect Christ to the story all the way through. In my mind I was doing this in a constructive way, but I am not convinced that the end result reflected an integrated and constructive sense of why Christ is significant throughout the biblical narrative. Having reflected on these matters more, I believe I can now show even more comprehensively how Christ matters to the entire biblical narrative, beginning with creation.

---

[3]Bethany Hanke Hoang and Kristen Deede Johnson, *The Justice Calling: Where Passion Meets Perseverance* (Grand Rapids: Baker Academic, 2015).

[4]See Kristen Deede Johnson, *Theology, Political Theory, and Pluralism: Beyond Tolerance and Difference* (Cambridge: Cambridge University Press, 2007), 162-67.

CHRIST AND CREATION

What I'm trying to get at is this. In the New Testament we find passages like Colossians 1:15-20 (NRSV):

> He is the image of the invisible God, the firstborn of all creation; for in him all things in heaven and on earth were created, things visible and invisible, whether thrones or dominions or rulers or powers—all things have been created through him and for him. He himself is before all things, and in him all things hold together. He is the head of the body, the church; he is the beginning, the firstborn from the dead, so that he might come to have first place in everything. For in him all the fullness of God was pleased to dwell, and through him God was pleased to reconcile to himself all things, whether on earth or in heaven, by making peace through the blood of his cross.

Or this one from the prologue to John (1:1-4 NRSV): "In the beginning was the Word, and the Word was with God, and the Word was God. He was in the beginning with God. All things came into being through him, and without him not one thing came into being. What has come into being in him was life, and the life was the light of all people." These passages show us that we need to consider creation in light of Christ—because in Christ all things in heaven and on earth were created. In Christ all things hold together. All things have been created through Christ and for Christ. All things came into being through Christ. Without Christ not one thing came into being. According to God's revelation in Scripture, this ought to—this needs to—impact how we understand the creation story and all that follows.

But I have found that easier to say theoretically than to apply constructively. And I don't think I'm alone in that. Here I offer a sampling of thoughtful, present-day Christians writing about what we can learn from the creation story, beginning with a couple of people who have participated in Center for Pastor Theologian conferences.

In his helpful and thoughtful work on culture, Andy Crouch traces our calling to be creators and cultivators of culture throughout the story of Scripture. He begins in Genesis 1, concluding from the creation story that "human beings will be responsible for the creation in its totality."[5] He places

---

[5] Andy Crouch, *Culture Making: Recovering Our Creative Calling* (Downers Grove, IL: InterVarsity Press, 2008), 103.

a great deal of emphasis on what is meant by "image" and "likeness" within the creation narrative, acknowledging that "generations of readers have offered suggestions ranging from the exegetical to the fanciful."[6] In his understanding, rooted in what we see of God's character over and over again in the first creation story, to be made in God's image is to "reflect the creative character of their Maker."[7] As Crouch goes on to unpack several marks of what this might mean, his focus is almost entirely on what we learn from Genesis 1 itself. He then moves to Genesis 2, exploring the human calling to cultivate what God has created. All of this is thoughtful and important; my summary here is not meant to suggest otherwise. Further, Crouch has a whole chapter on Jesus as culture maker, in which he references the very Colossians passage cited above.[8] The point I want to make is that what we know about Jesus Christ does not influence his reading of Genesis 1 and 2 or his interpretation of what it means to be made in God's image, nor does he, in his chapter on Jesus as culture maker, reflect back on the original creation story in light of what we learn about God through Jesus.

In a more recent work, *Playing God*, Crouch continues his exploration of what it means to be made in God's image by looking at power. The image bearers we read about in the opening chapters of Genesis are commanded to develop their own power, to use their power to create and shape the environment so that it can flourish in keeping with God's vision. Crouch acknowledges that as Christians we have the great privilege of reading this opening story from the perspective of Christ, maintaining that in Christ we see "true image bearing, true flourishing and true power . . . restored."[9] But that "restored" language is a clue that the creation story itself is still setting the terms. Later in the book Crouch has a lovely section on Jesus the true Image Bearer, who through his incarnation, crucifixion, and resurrection restores true image bearing. Through this work of reconciliation and restoration, the way to "dignified, delighted image bearing" is "re-opened" to the rest of humankind.[10] This, in fact, is how he grounds the call to evangelism

---

[6]Crouch, *Culture Making*, 103.

[7]Crouch, *Culture Making*, 104.

[8]Crouch, *Culture Making*, 134-46.

[9]Andy Crouch, *Playing God: Redeeming the Gift of Power* (Downers Grove, IL: InterVarsity Press, 2013), 36; see also 34-36.

[10]Crouch, *Playing God*, 80-81.

and social justice. Evangelism can be understood as "restoring the image bearers' capacity for relationship and worship, where the true Creator God is named, known and blessed."[11] Justice is restoring the conditions that make image bearing possible, working to enable conditions "where every image bearer can experience full dignity and agency."[12] There is much to be learned from this approach, but note the order. The opening creation story is setting the terms into which Jesus Christ fits, rather than the other way around. Crouch helps us ask the question, While honoring the biblical story as it unfolds, how do we nevertheless think christologically about what precedes Christ temporally?

Crouch's approach is similar to that taken by James K. A. Smith. In *Desiring the Kingdom*, Smith describes humans as those "called to be renewed image bearers of God (Gen 1:27-28)—to take up and reembrace our creational vocation, now empowered by the Spirit to do so."[13] When we receive a call to worship on a Sunday morning, Smith understands this call to be "an echo of God's word that called humanity into being (Gen. 1:26-27); the call of God that brought creation into existence is echoed in God's call to worship that brings together a *new* creation (2 Cor. 5:17). And our calling as 'new creatures' in Christ is a restatement of Adam and Eve's calling: to be God's image bearers to and for the world."[14] Notice here how creation is setting the terms—the worship we undertake today in Christ echoes back to God's original creation, the new creation is understood in terms of the old creation, and the calling we have as new creatures is defined in terms of Adam and Eve's original calling.

The calling is related to the image of God in us, which is not a property but a task—the task or mission of ruling and caring for creation, cultivating it, being priestly ambassadors of creation mediating God's love and care for creation.[15] And Jesus, as Smith describes him, can in turn be understood in light of this original calling: "Jesus shows us what it looks like to undertake that creational mission of being God's image bearer to and for the world. . . . Thus

---

[11]Crouch, *Playing God*, 80.

[12]Crouch, *Playing God*, 81.

[13]James K. A. Smith, *Desiring the Kingdom: Worship, Worldview, and Cultural Formation* (Grand Rapids: Baker Academic, 2009), 162-63.

[14]Smith, *Desiring the Kingdom*, 163.

[15]Smith, *Desiring the Kingdom*, 163-64.

Jesus is our exemplar of what it looks like to fulfill the cultural mandate."[16] In this reading, Christ himself is understood in Genesis 1 terms. But what would it look like to also understand Genesis in light of Christ? To understand what it means to be image bearers by letting the One who came as the image of the invisible God reshape our very notion of being made in God's image?

Let's look quickly at another contemporary figure with a slightly different focus. Nicholas Wolterstorff is an important conversation partner when it comes to God, Scripture, and justice. Writing first of shalom in 1983, he presents shalom as a comprehensive vision that animates Scripture, noting that it is first articulated in the poetic and prophetic literature of the Old Testament and is carried through to the New Testament. He describes shalom in a fourfold way as a person at peace with God, self, fellow human beings, and nature. He associates shalom with right, harmonious relationships as well as enjoyment and delight in and through those relationships. And even though shalom goes beyond justice, being more comprehensive in scope, "in shalom, each person enjoys justice," which to Wolterstorff means that "each person enjoys his or her rights."[17] While Wolterstorff acknowledges that language of shalom continues into the New Testament, when he is exploring what shalom entails in those four different layers, he turns only to the Hebrew Scriptures. Those provide the landscape, and then Jesus is understood within that landscape as the Prince of Peace, the one who guides us into the way of peace (Is 9:6; Lk 1:79).

Ultimately, Wolterstorff understands shalom to be "both God's cause in the world and our human calling." This calling he describes as "both a cultural mandate and a liberation mandate," with the cultural mandate going back to creation as articulated by "the creation perspective of the Amsterdam school" and the liberation mandate coming from passages like Isaiah 58 as highlighted by liberation theologians. He believes the shalom perspective incorporates but goes beyond what these two schools offer.[18] In putting forward his perspective, he does not explicitly refer to Jesus Christ. He does not draw on what we might learn from Jesus to reshape or set the terms of

---

[16]Smith, *Desiring the Kingdom*, 164.

[17]Nicholas Wolterstorff, *Until Justice and Peace Embrace* (Grand Rapids: Eerdmans, 1983), 69.

[18]He associates the cultural mandate with "the pursuit of increased mastery of the world so as to enrich human life," and the liberation mandate with "a struggle for justice" (Wolterstorff, *Until Justice and Peace Embrace*, 72).

understanding either our calling to engage and shape the world or our calling to liberate—despite the ways Jesus inhabits these roles so differently than we ever would or could have imagined on our own.[19]

I have learned from each of these writers. I admire each one as a person and thinker. I have assigned their writings in class and regularly lecture and teach from their works as I speak. But I still find myself wondering, What would shift if we moved from Christ to creation, from incarnation to image, from covenant to culture, from salvation to shalom, rather than always moving in the other direction? To put it differently, what is the cost of not looking both ways as we seek to understand these aspects of the biblical story and what they mean for our callings today?

CREATION DISTORTED?

As we ask these questions, we are in good company. Irenaeus, considered by some to be the church's first theologian, believed that we could not properly understand either creation or what it means to be human apart from Jesus Christ. Genesis alone will not suffice if we want a full picture of creation and humanity. This is why Irenaeus appeals to John 1:3 again and again as the key text when it comes to exploring the creation of humanity.[20] As Irenaeus writes, "The creator of the world is truly the Word of God. This is our Lord, who in these last times was made human. Existing now in this world unseen, he contains all things created and is immanent throughout the entire creation, for the Word of God governs and arranges all things."[21]

Irenaeus was concerned that trying to understand texts such as the opening chapters of Genesis apart from the revelation of God in Christ leads to heresy and discord.[22] This is what makes it so important to read creation in light of what Scripture reveals about Christ. As M. C. Steenberg notes,

---

[19]This assessment is based on a book written some time ago, but Wolterstorff's framework does not change much in his more recent books on justice. Indeed, he states in the first of these recent books that the book is his attempt to articulate the account of justice that he took for granted in this older book. See Nicholas Wolterstorff, *Justice: Rights and Wrongs* (Princeton, NJ: Princeton University Press, 2008), ix.

[20]On this point, see M. C. Steenberg, *Of God and Man: Theology as Anthropology from Irenaeus to Athanasius* (London: T&T Clark, 2009), 27.

[21]Irenaeus, *Against Heresies* 5.18.3, in *Irenaeus on the Christian Faith: A Condensation of "Against Heresies,"* trans. James R. Payton Jr. (Cambridge: James Clarke, 2012).

[22]See Steenberg, *Of God and Man*, 24.

Irenaeus believed that the wisdom found in such places as the prologue of John "offers the clarity needed by the Christian to unlock what otherwise might remain only a partial truth contained in the ambiguous anthropological statements in Genesis."[23] In other words, apart from Christ we simply do not have enough insight to unpack the opening statements found in Genesis related to what it means to be human, including what it means to be made in the image of God. Jesus Christ opens ways to see more fully what there, in those opening chapters of Scripture, we see only in part. In Irenaeus's words:

> For in long times past, it was said that humanity was created after the image of God. That could not be seen, though, for the Word after whose image humanity had been created remained invisible, and so humankind easily lost the similitude. However, when the Word of God became flesh, he confirmed both of these: he showed forth the image truly, since he himself became what was his image, and he re-established the image after a sure manner, by assimilating humanity to the invisible Father by the visible word.[24]

To put this more simply, when considering what it means that we humans are made in the image of God, why wouldn't we look at Jesus Christ, the one who "showed forth the image truly"? If Jesus is "the image of the invisible God" (Col 1:15), why would we not factor Jesus in as we seek to understand what it means that we humans are made in the image and likeness of God? Looking back to those opening chapters of Genesis in light of Christ only makes sense.

Many centuries later, Dietrich Bonhoeffer had similar convictions about the centrality of Jesus Christ to our understanding of Genesis. He writes, "The church of Holy Scripture—and there is no other 'church'—lives from the end. Therefore it reads the whole of Holy Scripture as the book of the end, of the new, of Christ. Where Holy Scripture, upon which the church of Christ stands, speaks of creation, of the beginning, what else can it say other than that it is only from Christ that we can know what the beginning is?"[25]

---

[23]Steenberg, *Of God and Man*, 25.

[24]Irenaeus, *Against Heresies* 5.16.2.

[25]Dietrich Bonhoeffer, *Creation and Fall: A Theological Exposition of Genesis 1–3*, in *Dietrich Bonhoeffer Works*, vol. 3, ed. John W. De Gruchy, trans. Douglas Stephen Box (Minneapolis: Augsburg Fortress, 1997), 22.

In Christ, God has made known the intentions he had from the beginning; the telos is revealed, and it is in the light of this end that we can make sense of the beginning. This conviction is woven throughout Bonhoeffer's theological exposition of Genesis 1–3.

To consider but one brief example of what this looks like in practice, when exploring what the first two verses of Genesis teach us about the beginning of creation, Bonhoeffer finds it essential to consider the resurrection of Christ. He suggests that

> it is because we know of the resurrection that we know of God's creation in the beginning, of God's creating out of nothing. The dead Jesus Christ of Good Friday and the resurrected Lord of Easter Sunday—that is creation out of nothing, creation from the beginning. . . . Yet the one who is the beginning lives, destroys the nothing, and in his resurrection creates the new creation. By his resurrection we know about the creation. For had he not risen again, the Creator would be dead and would not be attested. On the other hand we know from the act of creation about God's power to rise up again, because God remains Lord [over nonbeing].[26]

Like Irenaeus, Bonhoeffer applies the conviction that what we know about Jesus Christ gives us even more knowledge about creation.

In the case of both Bonhoeffer and Irenaeus, their theological convictions about the significance of reading Genesis in light of Christ are connected to distortions they saw happening around them. Irenaeus considered it essential to combating Gnostic interpretations of Scripture that questioned the goodness and God-createdness of the physical creation. Bonhoeffer followed Karl Barth in believing that a christological interpretation of Genesis countered readings of Genesis favored by some German theologians in which "orders of creation" (specifically around the supremacy of the German race) were used to support Nazi ideologies.

The truth that both Bonhoeffer and Irenaeus are pointing to is that the opening chapters of Genesis, while full of insight on some levels, are vague in other ways. If you were to take the first three chapters of any novel and place them in the hands of different people to fill in the rest of the story, you would end up with innumerable versions of what the story means and how

---

[26]Bonhoeffer, *Creation and Fall*, 35-36.

it unfolds. The story of Scripture is clearly not a novel, but when we treat those first three chapters without considering what follows, we end up with many different notions of what they entail.[27] Furthermore, the story of Scripture, unlike a choose-your-own-adventure story, unfolds in particular ways with Jesus Christ as its most central character. Why would we not try to understand the beginning of the story in light of the end revealed in Jesus Christ? Why would we not think this would help us understand God's intentions for creation and humanity from the start?

When we read the creation narratives on their own, it becomes all too easy to read into those opening chapters of Genesis widely divergent notions of God's intentions for creation, humanity, culture, or politics that come more from our context than from God himself. But when we begin with Christ in our reading of Scripture, we minimize the temptation to appeal to, as Alan Torrance puts it, "'foreign' categories and notions affirmed independently and in advance of God's Word to humankind." This helps us avoid the perennial Christian temptation to read into creation "such inherently risk-laden notions as the following: 'orders of creation'; naturalistic anthropological categories; individualistic theories of God-given rights; divinely ordained legal decrees and categories; cultural mandates; indigenous self-understandings."[28]

Why would Torrance consider such categories "inherently risk-laden"? Torrance, as a theologian in the Reformed tradition, is especially sensitive to the ways Reformed tradition had been co-opted and used to support tremendous political oppression in both Nazi Germany and apartheid South Africa. We may already be familiar with the ways German Christianity allowed itself to be co-opted by the convictions of National Socialism. We may be less familiar with the ways certain Dutch neo-Calvinists of South Africa were deeply influenced by the Reformed tradition's emphasis on creation and understood the segregation of the races to be a reflection of orders of creation established in Genesis. Salvation, understood in terms of the restoration of the order of creation, meant that humans living in

---

[27]Perhaps this helps explain why we have had so many different versions of what it means to be made in the image of God throughout the centuries.

[28]Alan J. Torrance, "Introductory Essay," in Eberhard Jüngel, *Christ, Justice, and Peace: Toward a Theology of the State*, trans. D. Bruce Hamill and Alan J. Torrance (Edinburgh: T&T Clark, 1992), xix.

Christ now were to be obedient to the mandates found within creation, including segregation.[29]

This might sound alarmist, but Torrance believes that operating with general concepts like orders of creation and cultural mandates makes it much easier to read cultural and political categories into those concepts than if we were to read creation through the lens of Jesus Christ. Barth shared a similar conviction. As Stephanie Mar Brettmann writes,

> For Barth, knowledge of good ethical action finds no basis in nature or culture or reason. In his own experience, such appeals led to disastrous consequences in Nazi Germany and compromised theology. For Barth, the answer to the question of justice is only perceived in the creature's acknowledgement of the moral field in which he has been set by God's encounter with humanity, in the person of Jesus Christ.[30]

If, as Hauerwas and MacIntyre remind us, notions of justice come from different and often incompatible sources; if, as Bonhoeffer, Barth, and Torrance remind us, Christian notions of justice often get distorted by cultural and political categories when they are not derived christologically; if, as Irenaeus reminds us, aspects of Genesis on their own are ambiguous and Jesus Christ was given that we might better know both God and humanity, then why would we not want to root our explorations of creation, humanity, and justice in Christ?

WHO WE ARE AS NEW CREATIONS

What do we learn if we take this christological approach to exploring creation and who we are as God's creatures? How might we better understand our callings to justice today if we look at them in light of Christ from the very beginning of the story? To help us explore this, we will engage with a handful of Christians who have asked similar questions.

Driven by the theological conviction that we need to take Jesus Christ as our starting point for exploring what it means to be human, contemporary

---

[29]See Eugene M. Klaaren, "Creation and Apartheid: South African Theology Since 1948," in *Christianity in South Africa*, ed. Richard Elphick and Rodney Davenport (Berkeley: University of California Press, 1997), 372-73.

[30]Stephanie Mar Brettmann, *Theories of Justice: A Dialogue with Karol Wojtyla/Pope John Paul II and Karl Barth* (Eugene, OR: Pickwick, 2015), 130.

theologian Marc Cortez asks us to consider what would happen if we took Pilate's claim upon presenting Jesus, "Here is the man"—*ecce homo*—as John's portrayal Jesus of as "the true human who comes to inaugurate the reality of new creation. As the *anthrōpos*, Jesus is the one who fulfills God's creational purposes for humanity."[31] This could be taken as creation setting the terms of humanity, but Cortez goes further: "If we say that Jesus is the true telos of humanity, the eschatological end that God had in mind from the beginning, it seems to follow that we must also maintain that this telos is intrinsic to the meaning of humanity. In other words, we cannot fully understand what it means to be human until we have seen the full humanity revealed in Jesus."[32] Cortez makes these claims in a fascinating study of the book of John's prominent theological motif of creation and new creation, in which Jesus is understood as one who both precedes and transcends Adam.

When considering creation itself, we see that Jesus did not come just to recapitulate the original creation but "to take the story in new directions, fulfilling the work of creation in a way that transcends what we had before."[33] This is another reason why we need to consider creation in light of Christ, because on its own the creation accounts do not show us the fullness of what God intended for his creation. Take John 20 as one example, in which we see the outpouring of the Holy Spirit after Jesus' death and resurrection: "Jesus said to them again, 'Peace be with you. As the Father has sent me, so I send you.' When he had said this, he breathed on them and said to them, 'Receive the Holy Spirit'" (Jn 20:21-22 NRSV), Cortez notes that this is an intentional allusion to Genesis 2:7 ("then the Lord God formed man from the dust of the ground, and breathed into his nostrils the breath of life; and the man became a living being" [NRSV]). With this pronouncement of shalom and the giving of the Spirit, we see the beginning of a new creation that transcends what was originally available in creation, for now the Spirit's presence in God's people is permanent, and through the Spirit they—we—are able to participate in the Son's intimate union with the Father.

---

[31]Marc Cortez, *ReSourcing Theological Anthropology: A Constructive Account of Humanity in the Light of Christ* (Grand Rapids: Zondervan Academic, 2017), 36.
[32]Cortez, *ReSourcing*, 36.
[33]Cortez, *ReSourcing*, 46-47.

This relational dimension revealed in Christ is hugely significant—the revelation of God as Father, the relationship between the Father and the Son, the gift of the Spirit that invites us into that shared communion between the Father and the Son. By God's grace, these relational realities made known in Christ help us understand more fully who God is and who God invites his people to be through the unfolding of creation.

For Karl Barth, creation can only be understood in light of God's covenantal purposes—God's desire for covenant with humanity, made known most fully in Jesus Christ. When reflecting on what it means to believe in God as Creator, Barth insists that we connect this to God as Father. He appeals to the first article of the Apostles Creed, "I believe in God, the Father Almighty, Maker of heaven and earth," writing, "The God who created heaven and earth is God 'the Father,' i.e., the Father of Jesus Christ, who as such in eternal generation posits Himself in the Son by the Holy Spirit, and is not therefore in any sense posited from without or elsewhere. . . . As He cannot be the Creator except as the Father, He is not known at all unless He is known in this revelation of Himself."[34]

By looking through the lens of Jesus Christ, we can see that God the Father and God the Creator are one and the same, pointing us to the biblical truth that God's intention in creation was always deeply relational and covenantal. As John Webster succinctly puts it, "The creation is (and therefore is known as) that reality which God destines for fellowship with Jesus Christ."[35] Biblically, Barth looks to passages such as John 1, Colossians 1, and 1 John 2 to support his conviction that we must look at creation christologically, noting, "The meaning of all these passages can only be that Christ stands as God and with God before and above the beginning of all things brought into being at the creation; He is the beginning as God Himself is the beginning."[36]

For Barth this means that we do not look at creation as it stands on its own and try to determine what it has to tell us about God's purposes or the nature of humanity. It's not the original order of creation that matters as much as its telos, which unfolds in the history of redemption found

---

[34]Karl Barth, *Church Dogmatics* III/1, ed. G. W. Bromiley and T. F. Torrance (Edinburgh: T&T Clark, 1957), 11-12. Hereafter *CD*.
[35]John Webster, *Karl Barth* (London: Continuum, 2000), 98.
[36]*CD* III/1, 51.

throughout the pages of the Bible.[37] This conviction is not meant to diminish the significance of the Old Testament. Barth still maintains that the church "cannot possibly overestimate the witness of the Old Testament. It will not be afraid, but rejoice, to allow the Old Testament history of creation to speak as a true and time-fulfilling history of the acts of God the Creator."[38] And yet that does not mean the Old Testament is sufficient to communicate on its own what it means for us to be human. For that we need to look at the person of Jesus Christ; "all anthropology should be based on Christology and not the reverse."[39]

That is to say, Jesus Christ defines what it means to be human. It's not that Jesus fits into a pre-existing category called "humanity." As Barth writes, "It is not the case, however, that He must partake of humanity. On the contrary, humanity must partake of Him. . . . As the nature of Jesus, human nature with all its possibilities is not a presupposition which is valid for Him too and controls and explains Him, but His being as a man is as such that which posits and therefore reveals and explains human nature with all its possibilities."[40] In terms of the image of God, Jeffrey McSwain argues, "On Barth's view, it does little good if we speak of humanity being created in the image of God unless we understand Jesus Christ to comprehend each anthropological aspect."[41] This is because Jesus Christ is uniquely the *imago Dei* (2 Cor 4:4). Further, Jesus Christ, as the firstborn of all creation, can be considered the original Adam (Col 1:15).[42] In short, to explore what it means to be human, we must consider Jesus Christ.

Irenaeus similarly argues that Christ's humanity, rather than Adam's humanity, is primary in our consideration of what it means to be human. As Gerald Hiestand writes, "For Irenaeus, Christ's incarnation forms the pattern for humanity's creation, rather than the reverse. . . . Irenaeus insists that Adam's humanity is made according to the image of Christ's humanity, not

---

[37]On this point, see Webster, *Karl Barth*, 63-64.

[38]*CD* III/1, 64.

[39]*CD* III/2, 46.

[40]*CD* III/2, 59.

[41]Jeffrey W. McSwain, *"Simul" Sanctification: Barth's Hidden Vision for Human Transformation* (Eugene, OR: Pickwick, forthcoming).

[42]See McSwain, *"Simul" Sanctification*.

the other way around."[43] What we see when we consider Christ's humanity is the centrality of the Father-Son relationship, in which the incarnate Son not only is a man born of Mary but exists in an active, ongoing relationship with the Father, receiving life from the Father in the Spirit.[44]

When we consider what it means for us to be made in the image of God, this relational dynamic is decisive. As Steenberg puts it, "To be created in the 'image' of this God-made-incarnate . . . is to be made an iconic representation of, and thus an active participant in, this full divine life of God as Father with his Son and Spirit."[45] This is to place our participation in the Son's communion with the Father by the Spirit at the center of God's ongoing desire for humanity. Jesus Christ is not primarily our example of how to live but the one by whose life, death, resurrection, and ascension we can finally participate in the relationship he shares with the Father through the Spirit.[46]

All of this suggests that the *who* is more significant than the *how* when we consider *why* we are called to seek justice in this world. At the heart of God's vision for creation is a people who are united to God in covenant fellowship.[47] By God's grace in Christ, we are at long last able to be God's covenant partners. It is from this place of union with Christ through the Spirit that we learn, from God and Scripture, who God is, who we are, and how we are called to live in this world.

For Barth, learning who we are includes recognizing that to be human is, first, to be with God and, second, to be for others. On the first point Barth writes, "Man is not without God, but with God."[48] Being with God is part of the very essence of what it means to be human. On the second point, when we look to Jesus Christ, we see that "the basic form of humanity, the determination of humanity, according to its creation, in the light of the humanity of Jesus—and it is of this that we speak—is a being of the one man with the other."[49] Summarizing this aspect of Barth's thought, Jeffrey Skaff writes, "While no one else saves others from sin, Christ reveals what is most

---

[43]Gerald Hiestand, "'Passing Beyond the Angels': The Interconnection Between Irenaeus' Account of the Devil and His Doctrine of Creation" (PhD diss., University of Reading, 2017), 75.

[44]See Steenberg, *Of God and Man*, 34-35.

[45]Steenberg, *Of God and Man*, 37.

[46]See Steenberg, *Of God and Man*, 45.

[47]With thanks to Jeffrey Skaff for the language of "covenant fellowship."

[48]*CD* III/2, 136.

[49]*CD* III/2, 244.

basic about the human's inner form, that human being is 'being in encounter' with other humans. . . . Barth beautifully elaborates what it means to exist as a 'being in encounter' with others: Humans must (1) look each other in the eye and be seen by another, (2) speak and listen to another, (3) assist one another, and (4) do all of this gladly."[50]

Writing more recently on the significance of the conviction that creation came into being through Christ, Norman Wirzba reaches a similar conclusion:

> To say that Jesus is the ontological principle governing all of creation is to say that his way of being, helpfully and succinctly described by Dietrich Bonhoeffer as his "being there for others," is determinative for the being of everything. To say that creatures come to be through Christ is to say that the existence and movement of each thing originates in and passes through the way of being-for-others that is embodied in Jesus' life.[51]

When we begin with Christ, we see that "being for others" is a defining part of what it means to be human. This way of life flows from who we are and is both made manifest and made possible through the life and sacrifice of Jesus Christ.

As Wirzba describes it, "When the eternal, filial, self-giving love of God became incarnate in Jesus of Nazareth, the whole of creation was introduced to and invited to participate in the hospitable ways of love that are the life of God and that have been operative all along."[52] Now, in the light of Christ, we can see more clearly the hospitable, self-giving ways of God throughout the biblical story, beginning in creation. We have become new creations in Christ by the grace of God, which means that we are able to participate in the Son's relationship with the Father thanks to the reconciling work of Christ, and then, as these new creations, we can receive God's calling to enter into the ministry of reconciliation in the world (2 Cor 5:16-21).

CONCLUSION

When it comes to talk of justice, it is easy to move right to discussions of what we ought to do to seek justice in this world. And it is easy to let our

---

[50]Jeffrey Skaff, "Theological Anthropology," in *Wiley Blackwell Companion to Karl Barth*, ed. George Hunsinger and Keith L. Johnson (Hoboken, NJ: Wiley Blackwell, forthcoming).

[51]Norman Wirzba, "Creation Through Christ," in *Christ and the Created Order: Perspectives from Theology, Philosophy, and Science,* ed. Andrew Torrance and Thomas McCall (Grand Rapids: Zondervan, 2018).

[52]Wirzba, "Creation Through Christ," 11.

notions of justice be shaped more by the latest political and cultural ideas than by what God has shown us to be just and right. While we want to be in conversation with people who have different notions of justice, as Christians we also want to intentionally root our call to justice in Christ and Scripture. Drawing from the perspectives of Christians throughout the centuries, from Irenaeus to contemporary theologians, who have been convinced of the centrality of Christ for reading all of Scripture, we see that this christological reading helps our notions of creation, humanity, and justice to be faithful to what God has made known about himself, his world, and us, his creatures. Why would we not understand God's purposes for humanity and creation in light of Jesus Christ, the image of the invisible God made visible in time and place for the salvation of the world? When we allow Christ to frame our understandings of creation, we see that our calling to seek justice flows from the invitation we have received in Christ to participate in the communion that Jesus the Son has always shared with the Father by the Spirit. With this framework in place, the invitation to seek first God's kingdom, justice, and righteousness is good news indeed.

12

Creation, Theology, and One Local Church in Southern California

GREGORY WAYBRIGHT

BECAUSE THE BIBLE'S OPENING WORDS DECLARE, "In the beginning God created . . ." (Gen 1:1), it is difficult for local churches to avoid questions related to faith and creation. In keeping with that, the Center for Pastor Theologians organized a conference titled "Creation + Doxology: The Beginning and End of God's Good World." This, of course, is no small topic. As Todd Wilson described it, "This conference will explore the cluster of topics related to the doctrine of creation, with a view to illuminating and applying the church's historic teaching on the beginning, purpose, and eschatological end of the world that God has made."[1] When I considered the amount of the Bible that touches on the topic of "the beginning, purpose, and eschatological end of the world," I realized that it is Genesis 1 through Revelation 22. I mentioned that to Pastor Carol Kenyon while she was the pastor to children at the church I serve and she immediately said, "Oh boy! That's my favorite part of the Bible!" I agreed.

The issue I was asked to address as a part of the conference was whether this topic is significant for the life of the local church and, if so, how we might approach it in our ministries. I imagine I was asked to do so partly

---

[1] Todd Wilson, "The Goodness of Creation," *Preaching Today*, September 2017, www.preachingtoday .com/sermons/sermons/2017/september/goodness-of-creation.html.

because of the kind of church I serve. The Lake Avenue Church (LAC) in Pasadena, California, is a 120-year-old urban church with a culturally and ethnically diverse congregation. Pasadena is the home of the California Institute of Technology (Caltech), including its affiliated Jet Propulsion Laboratory of the National Aeronautics and Space Administration (JPL/NASA). Because of that, our membership consists of many scientists, physicists, and engineers. At the same time, the Fuller Theological Seminary, one of the world's largest seminaries, is directly across the I-210 freeway from us. So, in a church with many scientists and seminarians, discussions about the relationship of this physical world to its Creator God are almost unavoidable. Still, because of the strong disagreements that often exist among church people about matters related to faith and science, we seem to have found ways to sidestep these discussions in church even though they are otherwise central to the lives of our people.

After I became senior pastor at LAC, I soon realized that if I care about my people's discipleship, I need to facilitate dialogue about science, creation, and re-creation. In this essay I seek to describe what I've learned. I first will tell stories of real people at LAC.[2] Then I will make a few broad concluding statements about how we might facilitate the kind of congregational culture conducive to people discussing issues like faith and science, issues that have all too often divided local churches.

True Tales of One Pastor Theologian's Parishioners

Dr. Steve Cunningham. Even though Steve's family sometimes went to church, they did not believe that God is personally involved in the world. Because of that, Steve grew up thinking that miracles don't happen. When he went to university, he made a conscious decision to reject the notion of God and to figure out for himself how to live. For many years Steve was educated as a scientist, a physicist, and as an engineer, earning his PhD in the field of theoretical solid state physics. Steve engaged in six years of postdoctoral work in related sciences, all confirming his thoroughgoing materialist worldview. This culminated in his teaching for three years at Caltech.

---

[2]All stories are told with permission.

After Caltech, Steve was hired by Hughes Aircraft and now works for the Boeing Company designing satellites and their systems. In 1984 he was selected to fly as a payload specialist on a space shuttle and would have flown in 1987—until, on January 28, 1986, the tenth *Challenger* shuttle exploded, killing the seven astronauts on board and ending both the program and Steve's hopes of being an astronaut. In the course of his studies, astronaut training, and professional work, Steve has published over fifty articles in scientific journals. He sometimes asks me to mention these things because he believes that what I report next might lead some people to think he is not a real scientist.

Steve's life continued on without any belief in God until things changed dramatically when his brother invited him to a meeting at a small church in Compton, California. At that church, Steve saw a blind woman instantly healed in response to a brief prayer, a woman whom he knew had been honestly and seriously blind for three years. That event left him wondering whether his worldview was too small to embrace all of reality. After wrestling with how such a healing could happen—surely with the prompting of God's Spirit—Dr. Steve Cunningham became a fully committed Jesus-follower. He still is.

How does a pastor theologian talk about both material and immaterial realities in a church that includes people like Steve Cunningham? I know Steve would want us to do so with courage and humility. I have found that Steve, and others like him, want their pastors to talk about faith and science with an unashamed conviction that God is involved in this world that he made. At the same time, he would have us do so without the kind of arrogance that suggests we know more than the Bible clearly teaches about questions like how and when God did what he did in creating. I believe Dr. Cunningham would not appreciate our telling him that he dare not grapple honestly with what he observes in his scientific work any more than he would appreciate his fellow scientists telling him that he must deny the miracles he sees in his walk with God. He also wants us to be a church community that is open to discovering more and more about the Word of God, the world God made, and the way God is involved in his creation.

As a final note, Dr. Steve Cunningham heads up the prayer ministries of the Lake Avenue Church and still gives testimony to experiencing miracles.

Min Li Shigematsu. Min Li was born and raised in China. Her childhood memories include stories of religious people being persecuted for being religious. Indeed, all religion was strictly forbidden in the society, especially for children. Religious books were banned. Marxism and Darwinism formed the intellectual foundation for what Min was taught in her schooling. But Min testifies, "Although I had never heard about God, God began to tug inside of me. All that I was taught left my heart unsettled. I knew something was missing."

At a young age, Min was burdened with "the causation question." She used to ask her grandma, "Who is my grandma's grandma's grandma . . . ? Who was the first grandma?" She had a deep intuition that there has to be some sort of uncaused cause in the universe.

Min also was affected by the reality of death. This became personal when both of her grandparents died prematurely during her teenage years. The pain of losing loved ones led her to face the problem of death and the meaning of life. Min says, "I struggled greatly knowing the truth that one day death will take all my loved ones away and it will take my life too. If everything ends in death, does life have ultimate meaning? With that question in my heart and mind, I was very unhappy."

Both of those issues—the cause of the universe and the reality of death—haunted Min into her university years. She asked about them unrelentingly. One day, after she had been discussing them with a friend and fellow student, the friend gave her a book and told her to read it. Min opened it at home and read, "In the beginning God created . . ." Min testifies, "The first sentence in the Bible brought my heart home and I felt peace."

She read the Bible voraciously and learned that her questions about death were also addressed—that death is not the end of things. Min accepted Jesus into her life and was baptized.

How would Min want me to talk about faith and science in church? I am sure she wants me to do so more often than I do! She is convinced that Genesis 1–3 is the key to the continuing growth of the church in China—with most people having been indoctrinated in communistic atheism or agnostic Buddhism. She points out that in those first three chapters of the Bible we discover the foundation for our understanding of who God is, who we are as human beings, and the world in which we live—both what is good about it and what has gone wrong.

Min does not want me to lose what she considers to be the main points she believes God makes in those chapters, that is, "I am. I am the Maker of all that is. I am ready to make myself known in this book."

As a final note, Min translates my sermons each week into Chinese, in addition to a series of messages in Genesis 1–3 that I recently delivered in China.

Dr. Lauren White. Lauren is an engineer, astrobiologist, and chemist at NASA's Jet Propulsion Laboratory. She is a true rocket scientist. I asked Lauren what she thought about how local churches might deal with matters of faith and science. Here is a part of what she wrote:

> I grew up in the South, where many churches subscribed to only one view of creation and the age of the earth. As I went down the path of becoming a scientist and ultimately an astrobiologist, it became more difficult for me to reconcile the church's views on creation with what I observed in the laboratory. For me, it has become a lifeline that Lake Avenue Church is willing to open the doors of our minds and to introduce us to many facets of thinking on creation that honors both the Bible and science. My eyes have been opened to be able to see that there is not only one way to interpret the biblical texts. I have realized I am not trapped between the rock of the Bible I believe and the hard place of the science I witness!
>
> It's helpful and encouraging to me in my faith as a scientist and engineer to know that I can do my work and not shy away from it in fear that I might somehow accidentally prove God was not involved. I have the privilege of being called to dig deeper and understand on the most foundational levels how God created and now sustains our universe. I am thankful that my church has been willing to bring people from different creationist platforms and hold special topic discussions about faith and science. When it does, it creates a safe environment for scientists who also know and love Jesus to discuss, on a scientific level, what their views are according to their research and to discuss without fear the struggles and questions they wrestle against. For me, my faith has only increased as I am able to do so. The further I investigate creation as a Jesus-following scientist and engineer, the more I see evidence of design and engineering and appreciate how God set in place the physics, chemistry, and biology to allow life to exist on our planet.

As a final note, Lauren is a contemporary hymn writer and one of the main worship leaders at our church. Her husband, John, is an engineer who has come to faith at Lake Avenue Church. Since his conversion, both he and Lauren have become some of our most fervent and effective witnesses to the gospel of Jesus Christ.

"Anne Onimous." I wish I did not have to tell you that Anne (not her real name) grew up in my church, although we all know of people like Anne in our churches. Anne grew up being very active in the church. This lasted all the way through her high school years. Her parents loved creation science. However, Anne was convinced that they would not allow their views about creation to be questioned.

Anne was and is intelligent. She received a scholarship to one of America's leading schools and then matriculated there. In her science classes, in her anthropology classes, in her psychology classes, and so on, she encountered perspectives about our world, views related to how cultures develop their beliefs, and perspectives on our human makeup that clashed with what she had been taught in her home and church. She had no tools by which to reconcile them. Eventually, she decided that what she had been taught to believe simply wasn't true. And she didn't have confidence that anyone in her church would genuinely be open to wrestling with her about the questions she had.

As a final note, Anne doesn't go to church much anymore—except occasionally on Christmas Eve, Easter, or Mother's Day.

I've been wondering what might help churches like mine to be meaningfully involved in the discipleship of people like Anne, Lauren, Min, and Steve. I believe that, as a part of my pastoral calling, I need to be a catalyst facilitating a congregational culture in which we are not afraid of difficult questions, even those that seem to pose challenges for our faith. Faith and science is just one area of discussion that we need to engage openly and without fear in our communities of faith.

What can we do to help the people in our churches navigate difficult issues like those related to faith and science, issues that often have divided local churches? I know we should not ignore them and hope they will go away. They will not! And I haven't found it to be helpful simply to make pronouncements from the pulpit. Instead, in my many years of experience as a pastor, I have come to believe that we must do four things.

NEED 1: CLARIFY THE CORE

When I have sensed a need for forums in the church to discuss controversial issues, I have learned that I need to enter the discussions with an understanding of what is at the core of biblical faith. In other words, as the apostle

Paul puts it in Romans 14, some issues are disputable matters in the life of a local church, and others are not. At LAC, we believe the non-disputable matters are summarized by the word *gospel*. We describe and communicate the doctrines central to the gospel by the way we have written our statement of faith.[3] In the preamble we write, "Our shared faith as members of the Lake Avenue Church family centers on God's evangel, the gospel of Jesus Christ. Through the power of this gospel, God accomplishes His salvation plan: rescuing His people from sin, making each one complete in Christ, and making all things in His creation new. Our most basic theological convictions are aspects of this gospel."

We will not budge from those "most basic theological convictions." That does not mean that those convictions are out of bounds for discussion. Far from it. We provide opportunities for people to engage in discussions about the articles in our statement of faith. However, our goal in those discussions is to defend these core convictions and to convince people of the truth of the gospel.

When it comes to disputable matters—and we believe that many questions related to the how and when of God's creating are among these—we seek to facilitate open and respect-filled dialogue. Such dialogue, in my experience, is only constructive for the life of the church when the church leadership has clarified the core convictions that provide the doctrinal basis for our unity.

NEED 2: PURSUE INTELLECTUAL RIGOR

After many years of being a pastor, I have discovered that the questions that have the potential to divide a church rarely deal with easy or simplistic issues. Therefore, I believe pastor theologians leading their congregations through questions related to faith and science need to be diligent in their own studies. I apply the apostle Paul's words in 2 Timothy 2:14-15 to these matters: "Keep reminding God's people of these things. Warn them before God against quarreling about words; it is of no value, and only ruins those who listen. Do your best to present yourself to God as one approved, a worker who does not need to be ashamed and who correctly handles the word of truth."

---

[3]Statement of Faith, Lake Avenue Church.

In particular, our people need us to help them discern what the issues are that the biblical texts call us to hold onto firmly and in what issues we might leave room for disagreement.

The longer I am a pastor, the more I find myself developing a muscle memory that approaches issues with a predetermined end. This leads to an intellectual rigidity that is lethal to good discussion. So I believe we need to avoid an approach that says, "This is where I have to end up, so how do I get there?" At the same time, I confess that I almost always enter into important discussions with personal convictions born out of years of study and reflection. The result of this is that, when I lead my people into discussions about challenging issues, I find I need to engage in a bit of a balancing act in the discussions. On one side, I need to listen to dissenting viewpoints carefully. On the other, I do not believe I should be reticent about defending my own beliefs. Engaging in both sides of that balancing act requires me to be a constant learner, continuing to study what scholars are saying about the topic we are discussing.

I was a seminary president for twelve years. In the course of that calling, I often sought to encourage the faculty by reminding them of the importance of their work—that is, that for the church's witness to the world to be effective, there is a need for scholars committed to the truthfulness of Scripture to continue to be involved in biblical/theological research and discovery at the highest levels. I am more convinced of that than ever. The demands of pastoral ministry often make it difficult for practicing pastors to devote the kind of time to scholarship that responding well to the ever-changing issues arising in our world seems to demand.

Having said that, I nevertheless believe that we who are pastor theologians also have an important role to play in bringing the truth of Scripture to bear on issues related to faith and science. In other words, our congregations need us to be pastor theologians. With regard to faith and science, there has long been a deeply rooted notion in many cultures that the material exists in opposition to the spiritual. This kind of dualism goes back at least as far as to the Manicheans, who believed the physical world was the creation of an evil god in constant conflict with a good god. The early church leaders had to deal with some congregants in their pews who saw the material world as inferior or even as evil.

As for us, ever since the emergence of modern science in the West, a significant group of people working in science has assumed that the material world and its processes preclude the spiritual. For generations now, both people of faith and people without faith speak of a sharp distinction between the physical and spiritual. The result is that many scientists view religious people with suspicion and many church people feel threatened by science.

Pastor theologians have a significant role to play in tearing down the dichotomy between the spiritual and material. We can do so by helping people navigate how to read and understand the biblical texts related to the material world and, specifically, to human origins. Without pretending to be scientists ourselves, we should be as informed as we can be about the issues emerging from scientific research that touch on our faith. We should also be the best exegetes and theologians that we can be. In this way, we can grow in the ability to facilitate the building of the kinds of intellectual bridges that enable mutual respect between the scientists in our churches and those who are not.

Bottom line: Pastors cannot simply leave all the scholarship to those in the academy. Pastors too must be as faithful to engage in scholarship as time will allow.

Need 3: Foster Personal Humility

Because Caltech and JPL/NASA are in my city, I occasionally have the privilege of meeting with groups of the scientists in my church. A few years ago, just before I was to moderate a Council for Christian Colleges and Universities (CCCU) forum on the teaching of faith and science, I met with eleven of our scientists. I asked them what the main thing was that they wanted me to pass on to those at the forum. They were unanimous in their response. They said that all respectable and honest scientists in all fields need to engage in their work with humility; that is, most scientists will admit that they do not yet know everything there is to know, even about their specific areas of expertise. They told me, "We all know that there have been times in our fields when people spoke with absolute certainty about one discovery only to have to moderate it later because of another discovery. We are all still learners."

My scientist-parishioners agreed that this is what I should pass on at the forum: "Tell your theologian and Bible-scholar friends that we will retain a measure of humility about our discoveries if they will do the same about theirs."

I sometimes watch, with occasional pain and embarrassment, videos of sermons I have preached. I see how easy it is for me to declare everything with passion and finality. I think that doing so is important when I speak about the core issues of our gospel-centered faith. However, I do not know everything. My people know this all too well. When I own up to that fact and admit it in a sermon, I find it to be freeing. Indeed, I discover how easy it is to call and get help from scientists in my congregation when I am struggling with a question touching on science. I've even detected that they are thrilled to be able to provide that help to their pastor.

When we engage in discussions from a mutual posture of humility, it's amazing how community-building, mind-stretching, and fruitful the discussions can be.

NEED 4: PRACTICE INTELLECTUAL HOSPITALITY

By intellectual hospitality, I am speaking of something like what Richard Mouw speaks of in his book *Uncommon Decency: Christian Civility in an Uncivil World.*[4] In addition to civility, I would like us to be communities of Christ-centered people who actually welcome ideas from dissenting voices. When we show hospitality, we welcome a person warmly and respectfully into our lives in such a way that we have the opportunity to shape one another's lives. Intellectual hospitality leads us to listen to the thoughts and perspectives even of those we disagree with. We listen without becoming defensive and without planning our argument against them even as they speak. I have become convinced that we need to listen carefully to the questions that people like my parishioner "Anne Onimous" are asking and then engage in a journey of seeking truth together.

This call to engage in intellectual hospitality is increasingly countercultural. On the one hand, many in our pluralistic culture seem to want to avoid any focused dialogue and debate about what is true. This kind of pluralism leads to preferring that others believe whatever they want to believe because,

---

[4]Richard Mouw, *Uncommon Decency: Christian Civility in an Uncivil World* (Downers Grove, IL: InterVarsity Press, 1992).

it is contended, truth is all subjectively determined anyway. On the other hand, those who are willing to enter into dialogue about what is and what is not true often do so in ways that parrot what we see in our politicians' debates in political campaigns. In those exchanges, there is no desire to listen to one another and to seek understanding together. Instead, opposing parties often seem only to want to win the argument.

Of course, we who, as the apostle Paul declared in Ephesians 2:1-10, were dead in our sins and have found grace through faith in Jesus should be able to relate to people differently. If we own the fact that we are recipients of unmerited grace, then we should be able to receive dissenting opinions more humbly and openly than others. However, this does not always seem to be what people experience in church. Dr. Michael Ressler, an astronomer and astrophysicist who is an active member at LAC, has put it this way: "We scientists often feel more comfortable talking about our faith with our scientific colleagues than we do talking about our scientific research with our fellow believers. We need to have forums where I and other believing scientists can take both very seriously and can do so with church members."

To show intellectual hospitality means we are desirous of considering an alternative position and also of challenging it. We engage in the discussion with an openness that hopes the other person's presence will also change us—that is, with the perspective that being with them will help us grow in ways that we could not if we spoke only with those we agree with.

This is the quality that made former Supreme Court justice Antonin Scalia able to gain much from those who disagreed strongly with him. He said, "If you can't disagree ardently with your colleagues about some issues of law and yet personally still be friends, get another job, for Pete's sake."[5] If that is true in the field of law, how much more so should it be true in theological discussions within the church?

We who are pastors and church leaders need to model intellectual hospitality to our people. I find it important to set up forums for faith and science to be discussed with respect and objectivity. Pastors can demonstrate intellectual hospitality in the way they participate in or moderate the forums. As

---

[5]Irin Carmon, "What Made the Friendship Between Scalia and Ginsburg Work," *Washington Post*, February 2016, www.washingtonpost.com/posteverything/wp/2016/02/13/what-made-scalia-and -ginsburgs-friendship-work/?noredirect=on&utm_term=.9d0f533c17c8.

a result, those who are employed in the natural sciences might experience that their church family believes that a person can be both a thoroughgoing follower of Jesus and a scientist. Ultimately, we pray that the way we as Jesus-followers interact about complex disputable matters might glorify God—that people who watch our interactions with one another may recognize the power of God to make peace even when controversial issues are being discussed.

Nurturing a culture in which issues of faith and science may be engaged with rigor and with humility is of great importance both to our churches' evangelistic witness and to our discipleship mission. This is summed up well by Dr. Lauren White: "The church that encourages its science- and engineering-minded followers to study and better understand the creation story in ways that include a scientific perspective as it relates to the biblical account is a church that strengthens followers in their faith and equips them to be a witness in possibly the toughest mission field there is—the intellectual field."

List of Contributors

Andy Crouch (MDiv, Boston University) is partner for theology and culture at Praxis, a community advancing redemptive entrepreneurship based in New York City. He is the author of *Culture Making*, *Playing God*, *Strong and Weak*, and *The Tech-Wise Family*.

Paige Comstock Cunningham (JD, Northwestern University School of Law; PhD, Trinity Evangelical Divinity School) is the executive director of the Center for Bioethics & Human Dignity, a Christian bioethics research center at Trinity International University. She has testified before Congress and is coauthor and editor of numerous book chapters, including "Evangelical Perspectives on Prenatal Testing."

Deborah B. Haarsma (PhD, Massachusetts Institute of Technology) is president of BioLogos. As an astrophysicist, she has published on gravitational lenses, young galaxies, and galaxy clusters. She is the author or editor of multiple articles and books on modern science and biblical faith, including *Origins: Christian Perspectives on Creation, Evolution, and Intelligent Design* and *Delight in Creation: Scientists Share Their Work with the Church*.

Gerald Hiestand (PhD, University of Reading) is senior associate pastor at Calvary Memorial Church in Oak Park, Illinois, and the cofounder and director of the Center for Pastor Theologians. He is the author and editor of numerous articles and books, including *The Pastor Theologian: Resurrecting an Ancient Vision* and *Beauty, Order, and Mystery: A Christian Vision of Human Sexuality*.

Kristen Deede Johnson (PhD, University of St. Andrews) is associate professor of theology and Christian formation at Western Theological Seminary in Holland, Michigan. In partnership with International Justice Mission, she and coauthor Bethany Hanke Hoang recently wrote the award-winning *The*

Justice Calling: Where Passion Meets Perseverance. Kristen's scholarship focuses on theology, culture, and political theory, and her publications include *Theology, Political Theory, and Pluralism: Beyond Tolerance and Difference*.

Michael LeFebvre (PhD, University of Aberdeen) is the pastor of Christ Church Reformed Presbyterian in Brownsburg, Indiana. He is the author of several books, including *Singing the Songs of Jesus: Revisiting the Psalms* and *Exploring Ecclesiastes: Joy That Perseveres*.

Hans Madueme (PhD, Trinity Evangelical Divinity School) is associate professor of theological studies at Covenant College in Lookout Mountain, Georgia, and an adjunct professor at Trinity Graduate School, Trinity International University. He is the coeditor of *Adam, the Fall, and Original Sin: Theological, Biblical, and Scientific Perspectives* and *Reading Christian Theology in the Protestant Tradition*.

Jeremy Mann (PhD candidate, Wheaton College) is a founder and the head of school at the Field School, a classical Christian elementary school on Chicago's west side that prioritizes students from low-income families.

John H. Walton (PhD, Hebrew Union College) is professor of Old Testament at Wheaton College. He has published nearly thirty books, among them commentaries, reference works, textbooks, scholarly monographs, and popular academic works. He is widely known for the Lost World books, including *The Lost World of Genesis One*, *The Lost World of Adam and Eve*, and *The Lost World of the Flood*.

Gregory Waybright (PhD, Marquette University) is the senior pastor of Lake Avenue Church in Pasadena, California. Prior to his ministry in Pasadena, Greg was president of Trinity International University in Deerfield, Illinois, for twelve years. He now serves on the Wheaton College Board of Trustees and is a frequent conference speaker with a focus on international leadership development.

Todd Wilson (PhD, Cambridge University) is the senior pastor of Calvary Memorial Church in Oak Park, Illinois, and the cofounder and chairman of the Center for Pastor Theologians. Todd is the author or editor of a number of books, including *Real Christian: Bearing the Marks of Authentic Faith* and *Mere Sexuality: Rediscovering the Christian Vision of Sexuality*.

Stephen Witmer (PhD, Cambridge University) is the lead pastor of Pepperell Christian Fellowship in Pepperell, Massachusetts. He teaches New Testament at Gordon-Conwell Theological Seminary and helps lead Small Town Summits, which partners with The Gospel Coalition New England to serve small-town churches and pastors. He is the author of *Eternity Changes Everything* and the volume on Revelation in Crossway's Knowing the Bible series.

Author Index

Subject Index

Scripture Index

CENTER FOR
PASTOR THEOLOGIANS

The Center for Pastor Theologians (CPT) is an evangelical organization dedicated to assisting pastors in the study and written production of biblical and theological scholarship for the ecclesial renewal of theology and the theological renewal of the church. The CPT believes that the contemporary bifurcation between the pastoral calling and theological formation has resulted in the loss of a distinctly ecclesial voice in contemporary theology. It seeks to resurrect this voice. Led by the conviction that pastors can—indeed must—once again serve as the church's most important theologians, it is the aim of the CPT to provide a context of theological engagement for those pastors who desire to make ongoing contributions to the wider theological and scholarly community for the renewal of both theology and the church.

Toward this end, the CPT focuses on three key initiatives: our pastor theologian fellowships, our *Bulletin of Ecclesial Theology*, and our annual conference. For more information, visit pastortheologians.com.

Finding the Textbook You Need

The IVP Academic Textbook Selector
is an online tool for instantly finding the IVP books
suitable for over 250 courses across 24 disciplines.

ivpacademic.com